P9-DIJ-075

Dr. Ecco's
Cyberpuzzles

Previous Books by Dennis Shasha

Out of Their Minds: The Lives and Discoveries of 15 Great Computer Scientists

Codes, Puzzles, and Conspiracy

The Puzzling Adventures of Dr. Ecco

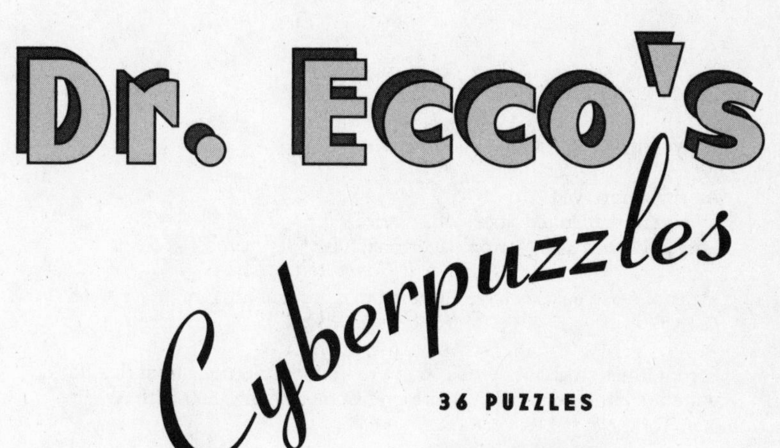

36 PUZZLES FOR HACKERS AND OTHER MATHEMATICAL DETECTIVES

Dennis E. Shasha

W. W. Norton & Company New York London

Printed in the United States of America
First published as a Norton paperback 2004

Many of these puzzles have appeared in a preliminary form in *Dr. Dobb's Journal*.

For information about permission to reproduce selections from this book, write to Permissions, W. W. Norton & Company, Inc., 500 Fifth Avenue, New York, NY 10110

Composition by Integre Technical Publishing Co., Inc.
Manufacturing by the Haddon Craftsmen, Inc.
Book design by Chris Welch
Production manager: Julia Druskin

Library of Congress Cataloging-in-Publication Data

Shasha, Dennis Elliott.
Dr. Ecco's cyberpuzzles : 35 puzzles for hackers and other mathematical detectives / Dennis E. Shasha.
p. cm.
ISBN 0-393-05120-X
1. Mathematical recreations. I. Title: Doctor Ecco's cyberpuzzles. II. Title.
QA95.S4693 2002
793.7′4.dc21 2003021869
ISBN 0-393-32541-5 pbk.

W. W. Norton & Company, Inc.
500 Fifth Avenue, New York, N.Y. 10110
www.wwnorton.com

W. W. Norton & Company Ltd.
Castle House, 75/76 Wells Street, London W1T 3QT

1 2 3 4 5 6 7 8 9 0

To the young ones:

Tyler, Jordan, Max, Jacob, David, and Caroline.

Contents

Contents

- ● Easy
- ■ Intermediate
- ♦ Hard

Acknowledgments

In January of 1998, Eugene Kim invited me to write a puzzle article for *Dr. Dobb's Journal*. On a lark, I replied that I'd be interested in recording Dr. Ecco's adventures as a column. Within 4 days, editor-in-chief Jon Erickson proposed an arrangement and, under the able stewardship of managing editor Deirdre Blake, we were off. The wonderful thing about the *Dr. Dobb's* readers is that they are high-quality programmers who love to think. This gave me the freedom to include Ecco puzzles that could benefit from computational power, while still demanding creativity and intelligence.

Many smart readers of *Dr. Dobb's Journal* helped improve the puzzles and the solutions. Some, but not all are acknowledged in the solutions that follow each puzzle. My thanks extend to every reader who sent me a solution. Many improved on Ecco's first cut.

As always, I have drafted my family in the solution of my puzzles. My wife Karen is my primary conscript, but my children Cloe and Tyler struggled with several of these puzzles, especially those parts aimed at cybernovices. Other family members also contributed, especially my brother Robert, my cousin Claire, and Dr. Ed.

Acknowledgments

Norton has been a wonderful publisher to work with. Editor Bob Weil and his assistant Jason Baskin made helpful suggestions with humor and insight. Production manager Julia Druskin kept me to schedule and ensured the highest quality result. Copy editor Carol Rose mixed thoroughness with good sense. Debra Morton Hoyt and Georgia Liebman deserve all credit for the design. Cover artist Matt Frost captured Ecco among his collage of thoughts. Justin Roth's drawings record some of the sinister locales.

Dr. Ecco's
Cyberpuzzles

Introduction

Is This Book for You?

Like so many mathematicians, Dr. Ecco loves puzzles. Fortunately for Ecco, people pay him to solve them. Joined by the narrator Professor Scarlet and Ecco's brilliant niece Liane, Ecco encounters a variety of puzzles posed to him by archaeologists, space station designers, generals, (reformed) criminals, and a few normal citizens. You are asked to solve them too. Some require new mathematical thinking. For most, the absolute best answers are not known, not even to Dr. Ecco. At the same time, these puzzles require no formal background beyond arithmetic and occasionally elementary algebra, though you might decide to augment your intelligence by using a computer.*

But don't worry. Even if you are not yet able to bend a programming language or spreadsheet in service to a solution, each puzzle has at least one cybernovice alternative. Alternatives so marked can be solved best by paper, pencil, and pure thought. Those marked cyberexpert usually require a technique of searching through a list of possibilities in some clever way. The best solutions to those variants are often still open.

Many of these puzzles appeared in preliminary form in *Dr. Dobb's Journal*, a magazine for professional programmers and other eccentrics. The readers of that journal have sent in many clever solutions

*When relevant, you can find the data for some of the puzzles at www.wwnorton.com/drecco.

to the "Omniheurist Puzzle Column" published there. When reader solutions are the best ones, I happily acknowledge their contributions.

Solving puzzles requires imagination, visualization, and experimentation. The ones in this book are intended to challenge and delight you. Enjoy them as you would a high-speed windsurfing jibe in the swells, a high-step dyno while bouldering, or a powder descent into a lightly forested ravine. Twist, think, and commit!

Hood River, Oregon

A Short Biography of Dr. Ecco

If Dr. Ecco had to identify his profession, he would call himself an *omniheurist*, a solver of all problems. Whether he means all problems or all kinds of problems, he has never told me. I can say only that I've never seen him fail.

Not that he doesn't have failings, though they're mostly of a moral nature. In this book, he consorts with smugglers and spies, an ex-con, and a tyrant. He turns away none of them. If a puzzle interests him, he will solve it.

How far from his humble beginnings in Chelsea, where I first saw him, a few years my junior, buying cake with his Mom. Sure he got the cake for free when he solved a brain teaser that would have stumped professional mathematicians. I thought maybe it was luck. It wasn't. After writing a brilliant doctoral thesis in mathematics while still in his teens, Ecco dropped out to pursue his present, well, profession.

In recording his adventures, I've tried to lay out the facts to you just as they are laid out to Ecco. You can match wits with him and his sassy niece Liane.

It won't be easy.

Warm regards,
Professor Justin Scarlet

P.S. I've violated the chronology in the puzzle ordering I present here. But that, as you'll see, is the subject of the last puzzle.

Part I

Design

1

Small Change for Mujikhistan

Ecco and I have a long-standing friendship nurtured by our frequent chess and Go games in his MacDougal Street apartment in Greenwich Village. "Eccentric" doesn't begin to describe his visitors.

"I manufacture money," Eli Tyler said, his ready grin making his eyes gleam. He was wearing a shirt with prints of beautiful coins from around the Earth. "I don't mean that I counterfeit it, but that I help governments design their coins. Here are a few of them." He pointed at five coins on his shirt. "I hope for some more. My problem is that the president of the new country Mujikhistan is a mathematician."

"What a terrible problem!" Ecco said, matching Tyler's grin.

"Yes, because he has this idea about saving money by minimizing the number of coins he needs to make," Tyler said. "He hasn't yet decided how many denominations of coins to mint, but he wants it to be the case that the average number of coins needed for a purchase should be as small as possible.

This adventure occurred in November 2000.

"For example, if the only denominations were pennies (each worth 1¢) and nickels (worth 5¢), then forming 99¢ would require 19 nickels and 4 pennies for a total of 23 coins altogether. The average number altogether would be the average of 1 [for 1¢], 2 [for 2¢], . . . , 2 [for 6¢], . . . , 6 [for 22¢], 7 [for 23¢], . . . , 23 [for 99¢]. That average is about 11.6. On the other hand, if the two denominations were pennies and dimes [worth 10¢], then forming 99¢ would require 9 dimes and 9 pennies for a total of 18 coins altogether. The average number of coins for forming 1 to 99 would then be about 9.1. So, dimes and pennies would be better, but then elves and pennies would also be better where an elf is worth 11¢. This president seems willing to consider any denominations at all that reduce costs to mint. 'If the denominations come out to be 37¢, 22¢, 4¢, and a penny, I would be happy. The people's mathematics level would improve,' he says. Can you imagine?"

"I bet you'd like this guy, Uncle," 12-year-old Liane said to Ecco.

"Except that he rolls heads when he's in a less numerate mood," Ecco said with a frown.

"These denominations should have some special properties," Tyler went on. "For example, we want it to be the case that the smallest number of coins can always (or nearly always) be obtained by counting out the money according to the following method: The first coin is of the largest denomination that is less than or equal to the amount desired; the second coin is of the largest denomination that is less than or equal to the amount desired less the value of the first coin; the third coin is of the largest denomination that is less than or equal to the amount desired less the value of the first two coins; and so on until the amount desired is reached. We call such a set *intuitive*. For example, the denominations 25¢, 5¢, and 1¢ are intuitive because it is never bad to start with 25¢ if the amount desired is over 25¢ and it is never bad to start with 5¢ if the amount desired is over 5¢ and under 25¢. On the other hand, if the denominations were 25¢, 10¢, and 1¢, then 30¢ would best be handled by three dimes, rather than a quarter and five pennies. Such a set of denominations is called *nonintuitive*. We are most interested in intuitive sets.

"To help him decide how many denominations to create, the president has asked me to find the best set of denominations that consists of (i) three coin values, including pennies; (ii) four coin values, including pennies; and (iii) five coin values, including pennies. He wants me to consider all prices between 1¢ and 99¢ to be equally likely and the goal is to minimize the average number of coins needed."

Liane and Ecco set to work. I went back to reading Robert Bly and was surprised that they came up with an answer in a little over an hour.

1. *Advanced cybernovice: Dr. Ecco was able to produce a three-denomination (including pennies) set that required on the average under 5.4 coins to pay for any item costing between 1¢ and 99¢; under 4.2 coins for a four-denomination set including pennies; under 3.6 for a five-denomination set including pennies; and under 3.3 for a six-denomination set. Can you do better? I'm sure you can.*

"What strange collections of denominations, hardly a 5 or 10 among them—and here you propose a 44¢ coin!" Tyler exclaimed when looking at the result. "The next generation in Mujikhistan will consist of bright people or calculator-toters. Tell me, is nonintuitive counting ever useful in your scheme?"

"Yes, I believe it's unavoidable," Ecco said. "The denomination sets we gave you are the best we could find using an intuitive counting method, but some of those sets are in fact nonintuitive, though only a few prices are best served by nonintuitive counting."

2. *Advanced cybernovice: In Dr. Ecco's case, the five- and six-denomination sets were nonintuitive. Dr. Ecco doesn't know whether his belief is true though. Can you resolve the question either way?*

"I'll have to think about that one," Tyler said. "But I have another question. I believe that with inflation, it will soon be the case that nothing will cost less than 5¢ and that, in fact, every item will cost a multiple of 5¢. In that case, could you design me the best coin values that are multiples of 5¢?"

3. *Cyberexpert: Dr. Ecco was able to design a three-denomination set requiring an average of less than 2.6 coins per purchase. For the best*

four-denomination set, the average was less than 2.2 coins per purchase; for the best five-denomination set, less than 2 coins per purchase. Can you do better? The results suggest that Tyler may have to design a 65¢ coin.

After Tyler left, Ecco turned to me. "Let's go back to the case where items can cost as little as a penny," Ecco said. "But now assume that the seller always has coins himself. So, for every price, we would like to minimize the sum of the number of coins passed from the buyer to the seller and the coins returned in change. We call this sum the *exchange number*. Notice that when exchanges are always possible, pennies themselves may not be necessary since an item costing a single penny may be purchased, if these coins are available, by paying with a 5 penny coin and receiving two 2 penny coins in return for a total exchange number of three. Assuming exchange numbers as the metric, what would then be the best three-denomination, four-denomination, five-denomination, and six-denomination sets, that is, the ones that minimize the average exchange numbers?"

I had no idea.

4. Cyberexpert: Would you like to attempt this one?

Solutions

I first present my original solutions and then improved ones suggested by readers.

1. Determining the best set of denominations for three, four, five, and six coin values, where each set must include pennies.

For three coins: 23¢, 5¢, and 1¢. This gives an average of 5.3131 coins per purchase for a number between 1¢ and 99¢.

For four coins: 38¢, 11¢, 3¢, and 1¢. This gives an average of 4.141 coins.

For five coins: 44¢, 17¢, 8¢, 3¢, and 1¢. This gives an average of 3.535354 coins using intuitive counting. The set, however, is nonintuitive (e.g., 24¢ is best served by three 8¢ coins).

For six coins: 45¢, 21¢, 8¢, 5¢, 2¢, 1¢. This gives an average of 3.242424 coins using intuitive counting. But this set is nonintuitive since, for example, 63¢ is best counted out by three 21¢ coins.

2. Ecco thinks that nonintuitive sets are necessary.

3. Here, all prices are to the nearest 5¢, and no pennies are allowed.

For a three-denomination set: 40¢, 15¢, and 5¢. The average number of coins needed is less than 2.53.

For a four-denomination set: 65¢, 25¢, 10¢, and 5¢. The average number of coins is less than 2.16.

For a five-denomination set: 65¢, 30¢, 25¢, 10¢, and 5¢. The average number of coins is less than 1.95 using intuitive counting.

4. The best overall solutions, in terms of completeness and original extensions, came from Ted Alper. Here were his solutions, some of which were equaled by other readers:

For three coins: If you insist on intuitive denominations, the best average you can get is 5.3131 . . . ; you can do this with denominations of 1¢, 5¢, and 22¢, but if you allow nonintuitive denominations, you can achieve 5.2020 . . . with denominations of 1¢, 12¢, and 19¢. And if you allow exchanges you can have an average exchange number of 3.9191 . . . with 16¢, 19¢, and 21¢ denominations.

For four coins: The best intuitive average is 4.1414, achieved with denominations of 1¢, 3¢, 11¢, and 37¢. The best nonintuitive average is 3.9293, achieved with denominations of 1¢, 5¢, 18¢, and 25¢. The best exchange average is 3.1212, achieved with denominations of 13¢, 28¢, 32¢, and 33¢.

For five coins: The best intuitive average is 3.4949, achieved with denominations of 1¢, 3¢, 7¢, 16¢, and 40¢. The best nonintuitive average is 3.3232, achieved with denominations of 1¢, 5¢, 16¢, 23¢, and 33¢. The best exchange average is 2.7273, achieved with denominations of 13¢, 24¢, 30¢, 32¢, and 33¢.

For six coins: The best intuitive average is 3.1616, achieved with denominations of 1¢, 2¢, 5¢, 11¢, 25¢, and 62¢. The best nonintuitive average is 2.9495, achieved with denominations of 1¢, 4¢, 6¢, 21¢, 30¢, and 37¢. The best exchange average of 2.5151 was achieved with denominations of 1¢, 8¢, 21¢, 38¢, 44¢, and 49¢.

He showed by comparison to the optimum possible that his solution is best: For 6 coins, exchanging for the values 1, . . . , 99: at most 6 values can be done with only 1 coin. You can get at most $(6 \times 5)/2 + 6 =$ 21 different possible values with 2 coins (no exchange) and at most another 15 values with exchanges of size 2 (bigger – smaller). So in the best case, in which *all* the remaining values take 3 coins, you'll have a total number of coins required of $1 \times 6 + 2 \times (21 + 15) + 3 \times (99 - 6 - 21 - 15) = 6 + 2 \times (36) + 3 \times (57) = 6 + 72 + 171 = 249$, which is achieved with the coin sets I already computed.

Here were Alper's solutions when each coin was worth a multiple of 5¢.

For three coins: The best intuitive average is 2.5263, achieved with denominations of 5¢, 15¢, and 40¢. The best nonintuitive average is 2.5263, achieved with denominations of 5¢, 15¢, and 40¢. That is, no nonintuitive set is better. The best exchange average is 2.2105, achieved with denominations of 5¢, 30¢, and 45¢.

For four coins: The best intuitive average is 2.1579, achieved with denominations of 5¢, 10¢, 25¢, and 60¢. The best nonintuitive average is 2.1053, achieved with denominations of 5¢, 10¢, 30¢, and 45¢. The best exchange average is 1.8947, achieved with denominations of 10¢, 25¢, 65¢, and 70¢.

For five coins: The best intuitive average is 1.9474, achieved with denominations of 5¢, 10¢, 15¢, 30¢, and 65¢. The best nonintuitive average is 1.8421, achieved with denominations of 5¢, 15¢, 25¢, 30¢, and 65¢. The best exchange average is 1.7368, achieved with denominations of 5¢, 10¢, 15¢, 45¢, and 80¢.

Several readers noted the obvious applications of a rational coinage. Martin Brown of Belgium observed: "It might interest you to know that the Belgian currency goes $\frac{1}{2}$, 1, 5, 20, 50, which as you can imagine leads to pockets full of change." Harm T. Voordenhout of the Netherlands thinks this question has continental applications: "I had been asking myself the same question about two years ago. The coming of the euro which would replace a lot of coins in Europe got me wondering about which system of coinage would be the best." It's too late now. Euro coins come in denominations of 1, 2, 5, 10, 20, and 50. The intuitive average is over 4.5. Alpert achieved 3.16.

Inspiration and Offshoots

I wrote this puzzle in the belief that countries must have figured out their coin denominations to minimize the number of coins needed to pay for goods. Little did I realize how naive I was. As simple experimentation and the puzzle solutions show, rather unexpected denominations would do a much better job. Even if one accepts the contention that most prices are a multiple of 5¢, the denominations are different from those of any country I know.

If the directors of the mint wanted to reduce the load on people's pockets, they might consider a statistical analysis of prices (with tax) and then perform the same optimizations that clever readers of this book might try.

2

Zoranic Monopoles

Architect and space station designer Jordan Tyler was back. "You helped us so much with our layout problem for the workspaces in the space station, Dr. Ecco," he explained as he entered, "that we've come to ask for your help in designing the cargo ports."

Ecco acknowledged the praise with a nod, then sweeping his arm in the direction of his 11-year-old niece, he said, "Liane played a significant role as well."

"True enough," Tyler said, smiling at Liane. "I'm glad to find both of you again. Here's our new problem. Monopoles, the cargo we're going to take to space, are unlike anything that's been taken before. They are constructed based on a physical principle that is still secret. I myself know only the effects.

"First, monopole triples of the proper polarities attract each other, much like the north and south sides of a magnet. Until a few decades ago, the very existence of monopoles was considered completely theoretical. It turns out, however, that there are many kinds of monopoles, in fact there are at least 52 of a special kind called *additive monopoles*. We know this, because Zoranics has manufactured them.

They claim that they are reaching a physical limit at 52, a limit that mathematical gambler Larry Laing has dubbed the "carddeck limit." I am skeptical that there is any such limit. Zoranics has given each kind of monopole a whole number label according to its properties: $1, 2, 3, 4, \ldots, 52$."

"What does the labeling mean exactly?" Liane asked, admitting the confusion we all felt.

"Good that you ask," Tyler answered with a smile. "Additive monopoles get their name because they exercise an enormous force of attraction if three of them labeled x, y, and z are in the same area and $x + y = z$. So, for example, the monopoles with labels 2, 5, and 7 will be attracted to one another. But 2, 5, and 6, for example, will exhibit neither attraction nor repulsion. Nor will 2, 3, 4, 9, because monopole attraction works only in triples. The attraction is so great that such a triple, once formed, is virtually impossible to break apart and is, therefore, useless to our mission.

"Fortunately the attractive force of the monopoles can be blocked by a shield of titanium. So, we are going to partition our cargo of monopoles into separate rooms with titanium shields to avoid creating mutually attractive triples. In each room, however, there must be no monopole triples x, y, z such that $x + y = z$.

"On the other hand, you can surely appreciate that we want to minimize the use of this precious metal. Thus our question: What is the smallest number of rooms you need if you want to send all 52 monopoles but only one instance of each? We'd actually like to send more, so if this is possible, please let us know."

"Wait," Liane interrupted. "I need an example."

"Fair enough," Tyler said. "Suppose you were allowed only two rooms A and B. Suppose you put the set of monopoles 1, 3 in room A and 2 in room B. Now, you can't put 4 in room A, because $1 + 3 = 4$, so you have to put 4 in room B. On the other hand, 5 can go into room A, so you would get 1, 3, 5 in room A and 2, 4 in room B. Unfortunately, you are now stuck. You cannot put 6 in either room."

Cybernovice: Can you find a strategy where you get stuck at 8 instead of 6?

Liane thought about this for a while and then said, "I see. Another strategy would be to put 1, 2, 4, 7 in room A and 3, 5, 6 in room B. Then you get stuck at 8 rather than 6. Still, we are far from 52. I wonder how many rooms we'll need."

Liane and Ecco worked on this for a while. Liane came up with a solution to the first question.

1. Cybernovice: Her strategy used only four rooms. How would you match that?

Advanced cybernovice: Can you get more than 52 in your rooms (e.g., over 60)?

"Thank you," Tyler responded. "Now, I must explain that we are not quite done. The space station directorate wants us to take pairs of each kind of monopole so we have spares. He says the mission is just too important. Further, the spare of each monopole should be in a different room from the original. That way, if any single room becomes damaged, we can collect all necessary monopoles from other rooms. Is doubling the number of rooms to eight sufficient in that case?"

"Let me make sure I understand," Liane said. "With this restriction, if we had three rooms, the best we could do would be some configuration like this:

Room A: 2, 3, 4;

Room B: 1, 2, 4;

Room C: 1, 3.

Here 5 could be put in room C, but no other room could accommodate 5. Is that right?"

"Exactly," Tyler said. "It's important that no room contain two monopoles of the same label."

2. Cybernovice: Dr. Ecco and Liane think that eight rooms are enough when two additive monopoles with the same label must go into different rooms. Can you do that well?

Cyberexpert: Can you do better (e.g., fit more than 60 with their spares in seven rooms)?

Tyler was clearly unhappy with this answer. "We may be able to convince the directorate that damage to rooms is such a remote possibility that it is OK to put two additive monopoles with the same label in the same room," Tyler said. "How many rooms would you need in that case?"

"Just a minute," Liane said. "Let's do an example again. Suppose a room has monopoles 3, 3, 4. That is, two 3s and one 4. Now, if I want to put a monopole of 6, can I do that?"

"No," Tyler answered, "because $3 + 3 = 6$. The rules remain the same. No, $x + y = z$, even if $x = y$."

3. Cyberexpert: How many rooms do you need if spares may go in the same room as the original and there is at least a full set of spares (52)? How many monopoles can you handle in that case? (Hint: well over 100 in 5 rooms)

Solutions

1. If duplicates are not allowed, then an "imbalance" strategy can put 52 monopoles into four rooms. Such a strategy consists of putting a monopole into the room that already has the most monopoles. Here is the distribution:

Room A: 1, 2, 4, 7, 10, 13, 16, 19, 22, 25, 28, 31, 34, 37, 40, 43, 46, 49, 52;

Room B: 3, 5, 6, 12, 14, 21, 23, 30, 32, 41, 50;

Room C: 8, 9, 11, 15, 18, 44, 47, 51;

Room D: 17, 20, 24, 26, 27, 29, 33, 35, 36, 38, 39, 42, 45, 48.

However, better solutions are possible. Ernst Munter was the first to find a 66 monopole solution:

Room A: 24, 26, 27, 28, 29, 30, 31, 32, 33, 36, 37, 38, 39, 41, 42, 44, 45, 46, 47, 48, 49;

Room B: 9, 10, 12, 13, 14, 15, 17, 18, 20, 54, 55, 56, 57, 58, 59, 60, 61, 62;

Room C: 1, 2, 4, 8, 11, 16, 22, 25, 40, 43, 53, 66;

Room D: 3, 5, 6, 7, 19, 21, 23, 34, 35, 50, 51, 52, 63, 64, 65.

2. Ecco and Liane showed that eight rooms are sufficient if every spare monopole with value *V* had to go into a different room from the original monopole of value *V*. Just double the number of rooms shown above.

Bart Massey was able to fit 65 monopoles with spares into only seven rooms. Here was his design:

Room A: 1, 3, 5, 10, 14, 16, 18, 20, 22, 29, 31, 44, 48, 50, 52, 56;

Room B: 2, 8, 12, 13, 16, 27, 32, 33, 38, 42, 52;

Room C: 1, 3, 5, 7, 9, 11, 13, 15, 17, 19, 21, 23, 25, 29, 35, 37, 41, 43, 45, 47, 49, 53, 55;

Room D: 24, 28, 30, 31, 32, 33, 34, 35, 36, 37, 38, 39, 40, 42, 44, 45, 46, 48, 49, 50, 51, 53;

Room E: 4, 6, 8, 9, 11, 21, 23, 24, 26, 39, 40, 41, 56, 57, 58, 59;

Room F: 2, 6, 10, 18, 19, 22, 26, 27, 30, 34, 43, 47, 51, 54, 55, 58, 59;

Room G: 4, 7, 12, 14, 15, 17, 20, 25, 28, 36, 46, 54, 57.

3. In problem 3, spares were allowed to be in the same room as the original monopole, but the monopole constraint (no two monopoles, even ones with the same value, added together could equal the third) still held. Nobody gave a correct solution with fewer than five rooms. Tomas Rokicki was able to put 123 monopoles in those five rooms—a remarkable achievement. In his solution, each number appears twice in the same room.

Room A: 1, 3, 5, 7, 9, 11, 13, 15, 17, 19, 21, 23, 25, 27, 56, 58, 60, 62, 66, 68, 97, 99, 101, 103, 105, 107, 109, 111, 113, 115, 117, 119, 121, 123;

Room B: 2, 6, 10, 14, 18, 29, 33, 37, 40, 41, 45, 48, 49, 52, 71, 72, 75, 76, 79, 83, 84, 87, 91, 95, 106, 110, 114, 118, 122;

Room C: 4, 12, 20, 22, 30, 36, 38, 39, 54, 57, 67, 70, 80, 85, 86, 88, 94, 104, 112, 120;

Room D: 8, 26, 31, 32, 35, 44, 47, 50, 53, 59, 65, 74, 77, 89, 92, 93, 98, 102, 116;

Room E: 16, 24, 28, 34, 42, 43, 46, 51, 55, 61, 63, 64, 69, 73, 78, 81, 82, 90, 96, 100, 108.

Inspiration and Offshoots

This puzzle is inspired by a subfield of combinatorics called Ramsey Theory, which starts with questions about small numbers of objects (like the minimum number of people at a party such that either four know one another or three are total strangers). The physics of the puzzle is quite fictitious. To begin, a magnetic monopole has one unit of magnetic charge and currently accepted theory suggests they exist, though none has been observed. The rest followed from practical considerations. Space flights must have multiple copies of many items. Some are dangerous if they are put together. The natural question is: How many can be packed into a small space? An excellent book on Ramsey Theory is *Ramsey Theory*, 2nd ed., by Ronald L. Graham, Bruce L. Rothschild, Joel H. Spencer, John Wiley & Sons, 1990.

3

The High Price of Safe Blood

He wore a sports jacket and a tie, but a stethoscope dangled from his neck. Dr. Max Jacobs looked us over for a few seconds each. "I hope you aren't squeamish, young lady," he said to 11-year-old Liane. "We're going to talk about blood."

"Human or newt?" Liane asked with a grin.

Dr. Jacobs smiled, "A quick wit. We'll need that for the problem I'm going to pose to you, Dr. Ecco, and the professor. As you may know, hepatitis C is a horribly debilitating illness that can be passed by blood transfusion. The test for the presence of hepatitis C is called the ELISA test. It isn't accurate enough, giving us both false positives (good blood is thought to be bad) and false negatives (bad blood is thought to be good). There is an expensive but relatively precise technique based on the polymerase chain reaction (PCR). This technique is just precise enough that if we take a drop of blood from 50,000 different pints and as few as one pint is contaminated with hepatitis C, we will detect it.

"Our center receives 100,000 pints per day. Very close to 1% of those pints is contaminated with the disease. We would like to figure out which pints are contaminated by using no more than 20,000 tests. Each test takes 2 hours, however, and we want to decide the fate of each pint in 4 hours or less. Can you help us?"

"How accurate do we have to be?" asked Liane. "I mean if you don't care whether you throw out some good blood with the bad, then we need no tests at all—just throw out the whole lot."

"You have a future in health administration," Dr. Jacobs answered with a grin. "To start with, please assume that you can't throw out more than two good pints with every bad pint."

"Let's first try the case where each pint can be tested for hepatitis C with only one test [2 hours] before its fate must be decided," Ecco suggested.

Cybernovice: Before you read on, see if you can find a method that results in under 35,000 tests and doesn't entail throwing out more than 2,000 pints of good blood, assuming there are 1,000 pints of bad blood. The fate of each pint must be decided in approximately 2 hours.

"Good idea, Uncle," said Liane to Ecco. "Well, since there are 1,000 bad pints, and we allow 2,000 false positives, we divide the pints into 33,333 groups of three and one group of one. We test all the pints in each group together. One thousand groups of three will have a bad pint. For each such group, we throw out all the pints."

1. Cybernovice: In the situation in which 4 hours are allowed for testing and there are 1,000 bad pints out of 100,000 and 2,000 good pints may be thrown out, Liane was able to do the job with under 12,000 tests. How well can you do?

"What if no errors are permitted?"

2. Cybernovice: Liane was able to handle the 4-hour no-error case in 20,000 tests. Can you?

"So, we either throw away perfectly good blood, or we need that many tests," Dr. Jacobs said sadly. "Suppose our pretest screening improved a lot and we could ensure that exactly one pint was contaminated. What is the smallest number of tests you could do with one 2-hour set of testing and assuming you don't want to waste any of that good blood?"

3. Cybernovice: Liane was able to do this with just 17 tests, but each test requires samples from many pints.

"Exactly one bad pint is unreasonable. What if there could be a small number, say up to 10. You still don't want to waste any good blood."

4, Cybernovice, but hard: Liane had no particularly good method for this which was guaranteed to give an answer in 2 hours. Even for 4 hours, her best answer required 1,900 tests in the worst case. She then revised this to 733 and finally to 714. Can you do better?

Dr. Jacobs had taken notes in an illegible medical scrawl. He even wrote down Liane's last solution. "You have a brilliant niece, Ecco. My compliments," he said. He folded his piece of paper into his jacket pocket, shook hands with us, and left looking very satisfied. Ecco was not.

"Liane, think," he demanded of his niece. "The gap between 17 and 714 is huge, all for a handful of pints that could be bad. Surely you can close the gap. How well could you do in 4 hours for just two bad pints?"

5. Cybernovice, but hard: For the case of up to two bad pints, Liane had a solution requiring under 41 tests and just 4 hours.

"Although blood can only be stored for 3 to 7 weeks before it goes bad, suppose I could give you as long as you need to make this work as efficiently as possible."

I never heard either of them speak more of the generalization of this problem.

Solutions

Keep in mind that there are many different ways to solve these questions, but as for just solving the problem with a 4-hour constraint on testing, the following two answers are outstanding.

Recall that the basic setting of the puzzle was that there were 100,000 pints of blood, some of which might be contaminated with hepatitis C. A PCR test that takes 2 hours is available and it can detect even tiny amounts of hepatitis, so drops from many pints may be combined and the hepatitis may still be detected.

1. For the first part of the puzzle in which 1,000 pints are bad, 2,000 good pints may be thrown away, and 4 hours are available, we proceed as follows:

a. Partition the 100,000 into 5,882 groups of 17 and 1 of 6.

b. Do 5,883 tests—one on each group. The worst case is that 999 tests are bad. (1,000 bad batches would be a good thing, since that means there is exactly 1 bad in each group. These could be tested as follows. Express each member of the 17 in its binary expansion starting with 0 and ending with 16: 0 = 00000, 1 = 00001, ..., 16 = 10000. Group all of those with high-order bit 0, matching 0xxxx. Group those with next high-order bit 0 x0xxx, then next xx0xx, xxx0x, and xxxx0. Test those in parallel. Since there is exactly one that is bad, let us say it is 10010. So, 0xxxx would test good. x0xxx would test bad as would xx0xx and xxxx0. But xxx0x would test good. So we would need just 5,000 tests in the second 2 hours.)

c. In the case where there are 999 bad batches, the set of potentially bad pints is $999 \times 17 = 16,983$ pints. We then partition these into 5,661 groups of three and test each group. We throw away all the pints in each bad group. The total number of tests is $5,883 + 5,661 = 11,544$.

This strategy, devised by Magne Oestlyngen, was the most clean and concise technique by far.

2. If no good pints may be thrown away with the bad ones, then we can be sure of finding an answer with 20,000 tests as follows. We divide up the 100,000 pints into 10,000 groups of 10. In the worst case, 1,000 groups will have bad pints and we then test all the pints among those 10,000. If fewer groups are bad, then we will have fewer than 10,000 individual pints to test.

3. Consider numbering the pints in binary. We need 17 bits:

0 0000 0000 0000 0000
0 0000 0000 0000 0001
0 0000 0000 0000 0010

and so on. Now test all pints with a 1 in the high-order position, then all pints with a 1 in the secondmost high-order position, and so on. Finally, test all pints with a 1 in the low-order position. The bad pint is the one whose ith bit (starting from the high-order position) is 1 if the ith test found a bad pint and is 0 otherwise.

4. Here is a method that finds up to 10 bad ones in under 2,000 tests in 4 hours. Test 1,000 groups of 100 each. If 10 such groups are bad, then we know there is one pint in each group and we use binary search as above. Otherwise, we test the 9 or fewer groups pint by pint. That requires something less than 900 more tests.

But one can do better as Dennis Yelle pointed out. Arrange the 100,000 samples in a large, nearly square rectangle matrix containing 316 rows by 317 columns. (The 172 extra spaces can be filled with empty containers.) In the first 2 hours, do 633 tests as follows: Do 316 tests where each test uses every sample in one of the 316 rows. Do 317 tests where each test uses every sample in one of the 317 columns. (Note that $316 \times 317 > 100{,}000$.) If either 10 rows or 10 columns test bad, there is one bad in each and we can use binary search on each (requiring only 40 more tests). So the worst case is that 9 rows and 9 columns test bad, giving a total of 81 more to test. This gives a total of $633 + 81 = 714$. Elijah Pau, Ng Weng Leong, and Tomas Rokicki contributed to this.

5. Here is a method that finds up to two bad ones in 41 tests in 4 hours. Number the pints 0, . . . , 99999. Now, rewrite each number as an eight-digit number using base 5, 5, 4, 4, 4, 4, 4, 4 (i.e., a base 5 most significant digit, a base 5 nextmost, and base 4 all the way down). There are $5^2 \times 4^6 = 102{,}400$ possible eight-digit (5, 5, 4, 4, 4, 4, 4, 4) numbers so each pint gets a unique number. (If base 5, 5, 4, 4, 4, 4, 4, 4 doesn't make sense to you, then imagine labeling each pint with a unique ID where the first and second digits are in 0, . . . , 4, and the remaining 6 are within 0, . . . , 3.)

Now, for each digit position, group all pints with a particular value for that digit position and test. Thus, we will run a test for 0xxxxx, 1xxxxx, . . . , 4xxxxx, as well as xx0xxx, xx1xxx, . . . , xx3xxx, and so on. The total number of tests in this first round is (5 + 5 + 4 + 4 + 4 + 4 + 4 + 4) or 34 tests.

Now consider the tests for the first digit position. There were five tests. One or two of them returned a hit. Ditto for all the remaining digit positions.

If any digit position returned only a single hit, then we know both bad pints in that digit position have that value at that position.

In at least one digit position, and probably most of them, we had two hits. That means that we have two possible values for that digit position. How do we match up the corresponding values?

For instance, if in the high-order digit position, we saw that either 2 or 3 tests bad, and in next high-order position, either 1 or 3 tests bad, do the bad pints start with 21 and 33, or 23 and 31? We don't know, but now we know that there is one bad pint whose high-order position is a 2 and another whose high-order position is a 3. So, in the second 2 hours, we look for one pint starting with a 2 using the technique we tried above. To be concrete, suppose that from the first test, we have bad batches for the following values:

Position 0 (high-order): 2, 3;

Position 1 (second- to high-order): 1, 3;

Position 2: 0, 2;

Position 3: 3;

Position 4: 1, 3;

Position 5: 0, 1;

Position 6: 0, 2;

Position 7: 2, 3.

To test 2 in the high-order, we will test:

21xxxxxx,

2x0xxxxx,

2xxx1xxx,

2xxxx0xx,

2xxxxx0x,

2xxxxxx2.

We don't need to worry about position 3. This will tell us which of the pints beginning with 2 is bad in at most 7 tests for each bad one (6 in this example). So this is 41 tests.

Generalization (pointed out by Suzanne Lea): You have n pints of blood and at most b bads. Any $m \leq n$ pints can be sampled and tested together. The test will say either there is at least one bad one in the sample or there is none. You have a maximum of b bad pints. What is the most efficient way to test assuming you have as much time as you want? This must be guaranteed to work. If there are none, then you need never consider them again. If there are b bad disjoint groups then you can use binary search on each.

For further reference, see Ding-Zhu Du and Frank K. Hwang, *Combinatorial Group Testing and Its Applications*, 2nd ed., World Scientific, 1999, p. 336.

Inspiration and Offshoots

The inspiration for this was a visit to the local blood donor center. They always take an extra vial to test whether your blood is good enough to go into the blood bank. I was wondering how they tested the blood under the assumption that most blood is good blood. It certainly seemed possible to do much better than test every pint. Little did I know that a mathematically inclined soldier had wondered the same thing when his blood was tested for syphilis upon induction to the army. He worked out a solution and there is a literature on this problem. In any case, some of the ideas presented by the readers seem to be novel and may contribute to any situation in which screening is necessary. My colleague Mike Goodrich of Johns Hopkins believes that probabilistic algorithms (ones in which random numbers are used to make decisions within a program) can lead to vastly better results than found so far. This is still an open question.

4

Greed and Good Breeding

Ecco feels no resentment toward the idle rich. He regards them with bemusement, as he would any deadbeat. Nevertheless, when I came over to his MacDougal Street apartment one fine autumn day, I was surprised to find him poring over the asset-intoxicated obituary of one Pier One Gorman. The article barely spoke of the accomplishments of the merchant, concentrating instead on the houses, yachts, and stock holdings to be divided among his heirs.

"Not what I would have expected you to read, Ecco," I said. "Are you jealous or something?"

"Who wouldn't be?" Ecco responded with a grin. "My excuse is that I have a professional interest in Mr. Gorman. You see his three heirs Alice, Brad, and Carla are due here any minute. It seems they have a private matter to discuss. No, no, don't leave. Nothing is private from you, my dear professor. Nor from Liane, for that matter."

Eleven-year-old Liane was lying on the couch reading Carl Sagan's book *Contact*. I was going to ask her about it but the doorbell rang.

"I assume I'm the first to arrive?" the young man asked. He was wearing an ascot, black leather driving gloves, and reflective sunglasses.

"Yes, Brad," Ecco said. "Please have a seat."

"My sisters are always late," he said with a tone of self-satisfaction. Just then, the doorbell rang.

The young woman who walked in wore haute couture from pillbox hat to heels. "Carla, Carla Gorman is my name," she said. "You are Dr. Ecco, I presume. I do apologize for being late, but the helicopter could find no place close to land." She nodded to her brother, sat down, glanced quickly around the apartment, and, apparently finding nothing of interest, opened the latest issue of *Avenue*.

A few minutes later, the doorbell rang again. This young woman wore jeans, no makeup, and a plastic earring on her left ear. "My name is Alice Gorman," she said as she shook hands with Ecco. "I'm the black sheep of the family." She said this last sentence with good humor and proceeded to kiss her brother and sister with an affection that they more or less returned.

Brad cleared his throat and began, "We have a set of 30 family heirlooms. We don't know how to distribute them. We can each jump up and down and say how much we value each one, but Carla wants everything for herself, Alice wants everything to give away, and I want the result to be fair."

"You just want a new Ferrari," Carla said, rolling her eyes.

"Does each of you have an equal right to the heirlooms?" Ecco asked.

"Yes, we are co-equal inheritors," Brad replied. "There are no others."

"I suggest a private auction then," Ecco said.

Carla, apparently experienced from her many visits to Christie's, glanced up in interest. "How do you mean?"

> *Cybernovice: Before reading on, can you think of a fair auction protocol?*

Ecco went on, "One thing you can do is to give each person a budget of 1,000 currency units, which I'll call CUs from now on. Each person can then bid for the 30 objects and pay in CUs. At the end, the CUs are worthless, so each heir might as well use them all. During the bidding, however, if heir X wins the bid for an object *o*, then only heir X pays CUs; the others keep the CUs they would have used for *o* had they bought it."

"That's a nice idea," said Alice. "What do you say, dear siblings, shall we give it a try? I have a list of the objects numbered from 1 to 30. Each of us should put a CU amount next to each object in such a way that the total CU amount on all objects is 1,000."

Brad and Carla apparently found no objection to this, so they created the following table.

OBJECT	ALICE	BRAD	CARLA
1	71	203	21
2	4	29	33
3	131	72	62
4	9	28	42
5	2	16	0
6	17	10	17
7	7	6	49
8	56	1	45
9	80	4	27
10	36	2	30
11	41	26	5
12	3	18	5
13	45	22	11
14	49	35	52
15	15	22	101
16	14	7	36
17	34	31	47
18	4	87	24
19	42	13	1
20	42	158	32
21	30	4	20
22	11	59	39
23	75	9	11
24	10	33	44

OBJECT	ALICE	BRAD	CARLA
25	64	1	124
26	12	19	19
27	18	33	61
28	7	11	28
29	34	29	3
30	37	12	11

Ecco looked over the table carefully. "Great," he said. "It's good to see that you value the objects so differently. Now here are the rules.

"We imagine that a computer program simulates an auction for each object. For each auction, the program compares the bids and assigns the objects to the highest bidder at a price that is one more than that of the second highest bidder. If there is a tie, then the person who precedes in alphabetical order wins. Any of your CUs that are left over after a bid are assigned equally to the remaining objects you are bidding on, with any remainder going to the next object.

"Here is an example in which each person has a total of just 66 CUs and there are 3 objects:

OBJECT	ALICE	BRAD	CARLA
1	23	25	29
2	11	15	17
3	32	26	20

"Object 1 is the subject of the first bid. Carla gets it for 26 CUs. Why? Because Brad would have bid 25, but no higher, so Carla would have won at 26. This yields the following table. Notice that the 23 that Alice would have bid for object 1 are split between objects 2 and 3. Similarly for the 25 that Brad would have bid for object 1 and for the 3 that Carla saved by paying 26 instead of 29.

OBJECT	ALICE	BRAD	CARLA	
1	0	0	26	—actually spent, Carla won
2	23	28	19	
3	43	38	21	

"Brad wins the second bid and pays 24 [one more than Alice]. This gives:

OBJECT	ALICE	BRAD	CARLA	
1	0	0	26	—actually spent, Carla won
2	0	24	0	—actually spent, Brad won
3	66	42	40	

"Naturally, Alice wins the last bid and pays 43. Now, all CUs disappear. Who got the better end of the deal?

"Carla received an object that she valued originally at 29, Brad at 15, and Alice at 32. This gives a total value of 29 + 15 + 32 or 76. The maximum deviation in happiness falls between Brad and Alice at 17.

"Notice, however, that the object each heir receives can change if we change the order of the auction. For example, if the program processes the objects in the order 2, 1, 3, then the distribution of objects proceeds as follows:

OBJECT	ALICE	BRAD	CARLA	
2	0	0	16	—Carla wins object 2 for 16
1	29	33	30	
3	37	33	20	

OBJECT	ALICE	BRAD	CARLA	
2	0	0	16	—Carla wins object 2
1	0	31	0	—Brad wins object 1
3	66	34	50	

"Alice wins object 3. So, Alice has finished in the same way, but Brad and Carla have swapped positions and Brad is in fact happier with this order. Alice has received goods she values at 32, Brad at 25, and Carla at 17. This gives a total value of 74 and a maximum discrepancy of 15. So, this solution is more equitable, but yields a lower value. That is, there are two possible optimality criteria: maximize total value or maximize fairness so the difference between the happiest and the saddest is minimized.

"We seek an order that achieves as high a value as possible and another order that is as equitable as possible. If the objects are auctioned off in the order of their object order, then the total value is 1,670 and the deviation (from happiest to saddest person) is 139.

"Here is the list of the results of the first 10 bids for the given order."

OBJECT	ALICE	BRAD	CARLA
1	0	72	0
2	0	0	49
3	98	0	0
4	0	0	53
5	0	20	0
6	27	0	0
7	0	0	25
8	55	0	0
9	44	0	0
10	40	0	0

Liane was writing something down on paper. "But there are more than 200 trillion billion billion possible orderings," she said.

"I see that Sagan has had a certain influence," Ecco said. "Nevertheless, we must find a very good order."

Cyberexpert: Dr. Ecco and Liane were able to find one ordering that achieved a total value of 1,801 with a deviation of 64 and another solution achieving 1,780 with a deviation of just 1. Can you find as good or better orderings? Better would mean either a total value greater than 1,801 or a deviation of 1 or 0 with a total value greater than 1,780.

After some debate, Alice persuaded her siblings to choose the ordering that gave the small deviation. They agreed after she gave each a week on her yacht to make up for the fact that she was the one who won 594 with that ordering, whereas they each won only 593.

Solutions

The ordering with total value 1,801 and a deviation of 64 is obtained as follows:

OBJECT	ALICE	BRAD	CARLA
27	18	33	61
14	49	35	52
7	7	6	49
19	42	13	1
25	64	1	124
6	17	10	17
16	14	7	36
5	2	16	0
2	4	29	33
23	75	9	11
12	3	18	5
29	34	29	3
15	15	22	101
11	41	26	5
13	45	22	11
10	36	2	30
30	37	12	11
3	131	72	62
24	10	33	44
4	9	28	42
17	34	31	47
8	56	1	45
21	30	4	20
22	11	59	39
9	80	4	27
28	7	11	28
18	4	87	24
26	12	19	19
1	71	203	21
20	42	158	32

A deviation of 1 with highest value 1,780 is obtained as follows:

OBJECT	ALICE	BRAD	CARLA
6	17	10	17
18	4	87	24
21	30	4	20
1	71	203	21
30	37	12	11
8	56	1	45
28	7	11	28
13	45	22	11
11	41	26	5
16	14	7	36
15	15	22	101
10	36	2	30
3	131	72	62
25	64	1	124
4	9	28	42
14	49	35	52
29	34	29	3
24	10	33	44
26	12	19	19
2	4	29	33
22	11	59	39
27	18	33	61
5	2	16	0
9	80	4	27
23	75	9	11
12	3	18	5
7	7	6	49
19	42	13	1
20	42	158	32
17	34	31	47

Inspiration and Offshoots

In his *Theory of Justice,* John Rawls suggests (this is my nonspecialist reading of his work) that the following thought experiment would lead to a just society: Imagine that you are about to be born but you have all of your adult rationality. However, you do not know which body or which life situation you will be born into. Which kind of society would you like to be born in? One with castes? One with pure equality but low material standards? One with inequality but high material standards? Rawls calls this lack of knowledge your "veil of ignorance" so the design is from behind this veil. Dividing the inheritance in the case of a vague will has some of the features of the thought experiment. One wants to come up with a distribution that everyone will be as happy with as possible, a kind of utilitarian partition of the goods. Now, through the veil of ignorance would you want to maximize total happiness or maximize fairness?

5

Gambling for Dear Life

"Wall Street is about revenue streams, Dr. Ecco," Dr. Dexter Coll explained after polite but efficient introductions. "Options help us generate new kinds of streams. Mathematics, in turn, helps us ensure that those option streams yield us positive net income at low risk while providing a service for our customers."

Dr. Coll, a mathematical inventor of great reknown for his seminal work in finance, seemed out of place in Ecco's chaotic apartment. Oak paneling would have matched his Saville Row suit much better.

"The option we want to sell now is called a 'getting even' option," Dr. Coll went on. "I'm not at liberty to tell you the real context, so I will give you the problem as a story whose mathematical features are isomorphic to the problem I must solve.

"Suppose you have a friend, Fred, who is in a dangerous situation. He loves to gamble and has made the mistake of gambling with some very unsavory characters called, well, the Sharps. He wants to stop, but he doesn't want to lose a lot of money. He's afraid he'll be killed if he stops when the Sharps owe him a lot of money. The trouble is he doesn't remember how much money he's winning or losing, and even more surprisingly, whether he's winning or losing. He knows only that he's either up or down something between $100,000 and $400,000. The Sharps, as evil as they are, are very honest about accounting and will tell him whether his outstanding credit is within $5,000 of zero after every bet. If he's either winning more than $5,000 or losing more than $5,000, all they will tell him is that he's 'out of bounds.' We don't like the sound of that phrase 'out of bounds.'

"The problem is that we don't know how to get back in bounds. The Sharps won't let him just pay up now or, in case he's winning, just forgive them the debt. They insist on being correct accountants."

"What game do they play?" Ecco asked.

"They flip a fair coin for even money," said Dr. Coll.

"At what stakes?" Ecco's niece Liane asked, forever interested in gambling.

"Any integer multiple of $1,000," Dr. Coll replied.

"How can I help?" Ecco asked.

"I want you to find Fred a strategy that gets him in bounds with as few flips as possible," Dr. Coll said. "You can assume that the coin is fair and that Fred has enough money to make any size bet. Also, he will pay attention from now on to whether he wins or loses a given bet. So, he will know how much he is winning or losing relative to his current situation. Finally, he can ask the Sharps whether he's in bounds or not after each bet. They will tell him yes or no. They always tell the truth."

"So, you want the shortest expected time in a statistical sense?" Ecco asked.

"Yes, that's exactly right," Dr. Coll said.

"What do you make of it, Liane?" Ecco asked.

"Can't Fred run out of money if he bets enough?" she replied with a question.

"For now, don't worry about that. We have, I mean he has, tremendous resources," Dr. Coll said with a fleeting smile.

Everyone fell silent. Liane was flipping coins and keeping count.

"OK, well I can solve an easier problem," Liane said. "Suppose we know that Fred is ahead by, say, $130,000, then I can make him be in bounds after just two flips in an expected sense."

Cybernovice: Try to figure out how before reading on.

Liane went on: "We have him bet as much as he is ahead. If he wins, then have him bet $260,000. If he wins again, then $520,000, and so on. The first time he loses, he'll be in bounds. On the average, he'll be in bounds after two flips, assuming the coin is fair."

"Right," Dr. Coll said, "but in this case you don't know where he starts or even whether he's winning or losing."

"Yes, I know," Liane said with a sigh. Straightening up, she said, "Oh, there is one more thing, though. Suppose all I know is that Fred is ahead by something between $125,000 and $135,000. Fred can still use my strategy for $130,000. The first time Fred loses, he'll be in bounds."

Ecco looked up, smiled, and nodded but didn't say anything. Soon, though, he was working out some calculations. "Liane, I think you are on to something," he said. "Let's consider one other simplification. Suppose that he could be ahead by either $100,000, $110,000, or $120,000 or behind by those amounts and you want to determine, by betting, how much he was ahead or behind by. In this case, in bounds means exactly zero. Suppose further that you knew in advance you would win the first bet, the third bet, the fifth bet, etc. [every odd bet] and lose the others. How could you quickly determine what Fred's initial situation was?"

1. Cybernovice: Give this a try. Five bets should be enough.

"With the values you gave me and Liane's idea, Dr. Coll," he said, "I can design a strategy for Fred that will get him to within bounds [between positive and negative $5,000, inclusive] after under 40 flips on the average. Fred will also find out approximately how much he was ahead or behind by. About 1 in 1,000 times, it will take him more than 100 flips."

2. Cyberexpert: Can you achieve an expected bound that is as good or better? If so, how?

3. Cyberexpert: Also, how long does it take you, on the average, to get in bounds if the bounds are between positive and negative $1,000, inclusive. Dr. Ecco can do it in under 160 flips on the average.

Dr. Coll listened carefully, asked a few questions, then stood up and put out his hand. "We have not yet discussed your fee, Dr. Ecco, but let's just say that it might change your lifestyle if you wanted," Dr. Coll said as he and Ecco shook hands. Then he turned to Liane.

"Liane, your question about whether Fred could go broke was a good one," Dr. Coll said. "His total capital is $10 million. In the case where in bounds means plus or minus $5,000, how likely is it that Fred will go beyond his capital at least once in a run? What if his capital were $1 billion? We need to know the answers to both of those questions."

Liane shrugged her shoulders. "I don't know. I'll go program some experiments in K, using my uncle's suggestions." She found that the numbers were remarkably high.

4. Cyberexpert: How likely is it that Fred will go broke with a capital of $10 million before getting in bounds at the $5,000 level? How about $1 billion?

Solutions

1. For this simplification of the problem, Fred is at either −$100,000, −$110,000, −$120,000, $100,000, $110,000, or $120,000. Fred first bets $110,000. He will win (because he wins on odd bet numbers). If he is in bounds, then he was losing $110,000 to begin with, so he stops. Otherwise he bets $10,000. He will lose. If he is in bounds, then he was losing $100,000 to begin with, so he stops. Otherwise he bets $20,000. He will win. If he is in bounds, then he was losing $120,000 to begin with, so he stops. At this point, he is up by $120,000 from his original situation. If not in bounds, he bets $230,000. He will lose. If he is in bounds, then he was winning $110,000 to begin with, so he stops. Otherwise he bets $10,000. He will win. If he is in bounds, then he was winning $100,000 to begin with, so he stops. Finally, he could bet $20,000, but doesn't need to

since we know what will happen by process of elimination. He would lose and he would be in bounds. His original gain was $120,000.

2. The solution to the original problem uses Liane's observation. Fred could be winning by between $100,000 and $110,000, between $110,000 and $120,000, and so on up to between $390,000 and $400,000. Similarly, he could be losing by that much. So, the basic strategy works this way: Let Fred's initial (unknown) winning or losing position be denoted *p*. Try to flip so that Fred will reach *p*+$115,000, *p*−$115,000, *p*+$125,000, *p*−$125,000, and so on up to *p*+$395,000 and *p*−$395,000 or until the Sharps tell him that he is in bounds. For example, if he is down $212,000, then *p*+$215,000 will put him in bounds. This gives 60 positions to search for. Call these the targets. On the average, he will have to explore 30 of them before getting within bounds. By aiming for each target individually, one's expected time would be two flips per target. One can take advantage of the fact that even when Fred misses a target, he may hit another.

For example, if he first bets $175,000, he loses, and he is not in bounds, then *p* cannot have been anything between $170,000 and $180,000 (otherwise he would now be in bounds). Following this loss, suppose that Fred bets $10,000. If he wins and he is in bounds, then the initial value *p* was between $160,000 and $170,000 (excluding the high value). If he loses and he is in bounds, then the initial value *p* was between $180,000 and $190,000 (excluding the low value). Thus, one can eliminate many possibilities with just a few bets, no matter whether Fred wins or loses.

Using a variant of the reader Wesley Perkin's terminology (he was the first to find this), a *p* value which has not been tested is called a "live" value. A *p* value that lies midway between the two other live values is called a "rich" value (and the two values are called its "neighbors"). A value which is both live and rich is called "live/rich." So, the idea is to test a live/rich value range (like $170,000 to $180,000 in the example above) and then make bets that test its neighbors ($160,000 to $170,000 and/or $180,000 to $190,000). The net effect is that one can solve the $5,000 spread problem in under 40 flips on the average. About 1 in 1,000 times it will go over 100 flips.

3. When in bounds means within $1,000, the average number of flips is about 150 using the same general method but will rise above 400 about 1% of the time.

4. Under the original formulation (in bounds means within $5,000 of 0), Fred will go negative by more than $10 million about 10% of the time and will go negative by more than $1 billion under 0.5% of the time. Under the $1,000 formulation, Fred will go negative by more than $10 million about 30% of the time and will go negative by more than $1 billion about 1% of the time.

Inspiration and Offshoots

A common misunderstanding of the "law of averages" is that a coin flipping game in which each bet is for $1 will lead to a tie over the long term. In fact, the lead one player or another will have tends to increase with the square root of the number of flips. The challenge in this problem is how to get back to even money. As the bets continue, one can, in some cases, be heavily in debt to the Sharps. A problem I don't know how to solve is whether one can get back to even money with high probability if the Sharps require one never to exceed some amount of debt.

Part II

Combinatorial Geometry

6

Chimps for Ana Ban

"They will simply die if we don't help them, Dr. Ecco," Dr. Ana Ban said. The great naturalist stood there, jungle fatigues and all. Twelve-year-old Liane was in awe. The woman exuded grace and a sense of purpose.

"I've been working with these chimps for the last 8 years," she said. "They are unique specimens. But I'm in Africa in the first decade of the millenium and some greedy officer wants to replace the current corrupt dictator with his own cronies. Last time this happened, all the animals in the zoo were killed in the cross-fire. This time, I'm taking no chances. I've got a huge container on a cargo ship and the shipping company is willing to drill air holes in the container so the ship can take the chimps to relative safety in Madagascar. The container will house 27 cages in a 3 by 3 by 3 cube. We'll put one chimp in each cage. Some of the chimps love each other and quite a few hate one another. I'm not sure how to place them to avoid battles and to keep the loved ones together. Can you help me?"

"Yes," Liane said eagerly. "If it's possible, we'll do it. Tell us which chimps love which ones and which ones might fight which ones."

"Very well," Dr. Ban replied. "Please understand that we know our chimps very well and give them human names that match their temperaments. Some of the names are familiar from history and others are the names of our colleagues, some of whose personality traits we see in the chimps. Their names are: Annie, Aristotle, Augustine, Caesar, Carlie, Cecily, Chimpie, Coco, Cordelia, Darwin, Didi, Emily, Goodall, Humboldt, Jonathan, Julia, Katie, Lea, Marie, Miles, Millie, Napoleon, Natasha, Sarah, Theo, Victoria, and Zoe.

"The loves and hates of our chimps are symmetric, so if chimp A loves chimp B, then chimp B will love A. Similarly, for hates. If a chimp loves another one, then it should be next to that one on the same level of the container and they should share a wall. They will want to touch each other if the water turns rough. If a chimp hates and might fight with another one, it should not be next to that other one vertically or horizontally. Not even diagonally at a bottom or top corner. Their arms are long and their hands strong."

"So, the chimp in the center row of the middle level had better have no enemies," Liane pointed out.

"That's right," said Dr. Ban, looking at the young girl. "You may well help me, as you claim. Here is the description of the loves and hates of our chimps:

"Aristotle loves Lea.
Chimpie loves Marie and Cecily.
Darwin loves Annie and Millie.
Humboldt loves Zoe.
Goodall loves Jonathan.
Napoleon loves Emily.
Caesar loves Julia and Cordelia.
Theo loves Zoe.
Augustine loves Didi and Katie.
Coco loves Theo.
Didi loves Augustine.
Carlie loves Jonathan.
Aristotle might fight with Chimpie, Augustine, Annie, and Marie.

Chimpie might fight with Caesar, Cordelia, Katie, and Carlie.
Darwin might fight with Humboldt, Napoleon, Sarah, Victoria, Miles, and Emily.
Humboldt might fight with Natasha, Didi, Katie, Annie, and Marie.
Goodall might fight with Zoe, Sarah, Didi, Carlie, and Lea.
Napoleon might fight with Caesar, Millie, Natasha, Augustine, Julia, Didi, Lea, and Marie.
Caesar might fight with Natasha, Sarah, Lea, Marie, and Emily.
Zoe might fight with Millie, Natasha, Augustine, Didi, and Victoria.
Millie might fight with Cordelia, Didi, Carlie, Miles, and Lea.
Natasha might fight with Cordelia, Coco, Julia, and Emily.
Theo might fight with Augustine, Didi, Katie, Cecily, Lea, and Marie.
Augustine might fight with Cordelia, Coco, Sarah, and Emily.
Cordelia might fight with Coco, Julia, Katie, Miles, and Lea.
Coco might fight with Sarah, Didi, Katie, Carlie, Annie, and Lea.
Julia might fight with Sarah, Victoria, Carlie, Miles, Annie, and Emily.
Sarah might fight with Didi and Victoria.
Didi might fight with Victoria, Katie, and Emily.
Victoria might fight with Katie, Carlie, and Annie.
Katie might fight with Carlie, Miles, Cecily, Annie, and Emily.
Carlie might fight with Marie and Emily.
Miles might fight with Annie.
Cecily might fight with Annie and Emily.
Annie might fight with Marie and Emily.
Lea might fight with Marie and Emily.
Marie might fight with Emily."

1. Cybernovice: Can you find a placement of the chimps in a 3 by 3 by 3 cage container in which each chimp will share a wall with all the chimps it loves (a wall is better than a floor or ceiling) but will not share even a corner with a chimp it hates? You might map out your answers as three 3 by 3 matrices, each one indicating a floor.

Liane worked out a solution in a remarkably short 30 minutes. Dr. Ban took some time and then verified it against her list. "There is one more thing," she said with a sigh. "Chimps that hate one another shouldn't be in the same column of the container. The feces will drive the bottom chimps crazy even if they can't reach one another."

Liane managed to solve the problem even with this constraint.

2. *Cybernovice: Can you find a placement of the chimps in a 3 by 3 by 3 cage container in which each chimp will share a wall with all the chimps it loves (a wall is better than a floor or ceiling) but will share neither wall nor corner nor column with a chimp it hates?*

Dr. Ban thanked us all, but most especially Liane, and then left.

"Well done, Liane," Ecco said to his beloved niece. "Could you have solved it if, in addition to the latest rules of Dr. Ban, no two chimps who hate each other could be on the same level?"

"I have a movie to make, Uncle," Liane said as she rushed out the door with her digital camera. I never heard the answer.

3. *Cyberexpert: Would you like to take a crack at this? What is the minimum number of conflicting pairs who violate the additional rule?*

Solutions

1 and 2. Here is Liane's solution to the second problem Dr. Ban posed. This solution subsumes the answer to Dr. Ban's first one.

Top level:

Aristotle, Cecily, Chimpie
Lea, Natasha, Marie
Didi, Augustine, Katie

Next level:

Sarah, Miles, Victoria
Carlie, Jonathan, Goodall
Annie, Darwin, Millie

Bottom level:

Humboldt, Napoleon, Emily

Zoe, Theo, Coco

Cordelia, Caesar, Julia

Larry Lewis found 16 different (but more or less symmetrical) solutions to the problem after making his Pentium work for 12 hours. Mark Taylor agreed with this and showed that they are symmetric variations of one another based on reflections and rotations.

3. Exhaustive analysis (trying all possible configurations) shows that the third problem, "chimps who hate each other shouldn't be on the same level," cannot be solved. Patrick Schonfeld showed that the minimum number of conflicting pairs in such a situation is 24, while still solving the first two problems.

Inspiration and Offshoots

OK, I admit it. This one was inspired by observing kids during a class trip. Kids form alliances and establish dislikes. How could one place them so nobody has to be near someone he or she doesn't like? Kids were possible, but chimps were better. There were newspaper articles about assaults on the habitat of the great apes. These two together gave rise to the problem. The general mathematical problem is to find permutations of the nodes in a graph G1 such that nodes that are neighbors in some enemies graph G2 are not neighbors in G1.

7

The Secrets of Space

Great architects like Ecco. They recognize his ability to solve tough structural and placement problems. The feeling is mutual. Ecco doesn't care much for glass towers or phoney Victorian facades, but he holds a deep affection for Italian piazzas, well-designed airports, and clever approaches to traffic flow like those of the Guggenheim museum in New York or the Senkai complex in Tokyo. The kinds of problems he's attracted to, though, usually have to do with extreme environments such as Antarctica or underwater cities.

One breezy winter day Jordan Tyler brought him such a problem. Tyler's reputation had preceded him. A descendant of a famous railroad designer, he combined mathematics, architecture, and physics in all his creations. It was for that reason that a consortium of space exploration companies led by none other than Mr. Nog Tugget (readers may remember this unsavory descendant of a ruined Alaskan gold rusher) had approached Tyler to design a modular space station. Tyler's design specified rooms consisting of symmetric cubes 10 meters on a side with a single manhole-sized door in the center of each of the six cube faces.

"We decided there was no reason to prefer one direction to another in space," Tyler explained. "In fact, I'd like to get away from the idea of rooms altogether, but the oldtimers want to be able to quarantine areas in case of leaks or toxicity problems. They've also asked me to help them design the interiors of the 48 cubes, but the first problem is how to arrange them and which company to place in each cube. That's what I'm coming to you for."

"What have they told you about what goes on in the space station?" Ecco asked.

"It is a deep commercial secret, it seems," Tyler answered. "All I know is that 15 companies are involved, each of which occupies one or more cubes that must be connected. I'll indicate this to you with a letter for each corporation, because their names are also secret."

COMPANY	NUMBER OF ROOMS
A	10
B	1
C	8
D	2
E	13
F	2
G	2
H	1
I	1
J	1
K	1
L	2
M	2
N	1
O	1

"Also, certain pairs of companies, due to joint ventures, must have neighboring cubes. Here I represent the necessary adjacency relationships in alphabetical order. For example, the fourth row says that some room of D must be next to some room of A and some [possibly different] room of D must be next to some room of C, and so on for E, F, and G. There is some redundancy in this representation, but I hope it is clear.

"Lack of adjacency imposes no constraint, by the way. The fact that B is not in the adjacency set of D, for example, means that it is not necessary for any room of B to border any room of D, though this is allowed and even desirable. These companies engage in all manner of joint ventures and those joint ventures may require an exchange of materials within the space station."

COMPANY	NECESSARY ADJACENCY RELATIONSHIPS
A	B, C, D, E, F, H, I, J, L, M
B	A, C
C	A, B, D, E, G, M
D	A, C, E, F, G
E	A, C, D, F, G, J, K, L, M, N, O
F	A, D, E, J, N
G	C, D, E, K
H	A, I, L
I	A, H, J, L
J	A, E, F, I
K	E, G, O
L	A, E, H, I
M	A, C, E
N	E, F
O	E, K

"From your encoding, can one conclude that M is a card sharp and K is an empath?" asked Ecco with a smile.

"Random letters produce tantalizing patterns," Tyler answered with a chuckle.

"And when you say the 13 rooms of A must be connected, all you mean is that an astronaut must be able to float from any room in company A's space to any other without passing through any other company's cubes?" asked Liane.

"Correct," Tyler said, nodding to the 10 year old. "As I said, each company is very protective of its secrets, more secretive in many ways than even the military establishment."

Advanced cybernovice: Dr. Ecco and Liane succeeded in discovering a placement of the cubes in three dimensions and an assignment of the companies to cubes that satisfied these constraints. Can you do it?

After Tyler left, Ecco turned to me. "My dear professor," he said, "do you think it is possible to do better? I mean can you find a design that satisfies these constraints but that makes more companies adjacent? If so, what is the design and what is the largest number of adjacency relationships you can find?"

Cyberexpert: Would you like to give it a try? The best I've seen is given in the following connection diagram:

COMPANY	CONNECTIONS
A	B, C, D, E, F, G, H, I, J, K, L, M, O
B	A, C, E, J
C	A, B, D, E, F, G, H, I, J, M, N
D	A, C, E, F, G
E	A, B, C, D, F, G, H, I, J, K, L, M, N, O
F	A, C, D, E, J, K, N
G	A, C, D, E, K, M
H	A, E, I, L, O
I	A, E, H, J, K, L
J	A, B, E, F, I, N
K	A, E, F, G, I, O
L	A, E, H, I
M	A, C, E, G, O
N	C, E, F, J
O	A, E, H, K, M

Solutions

Ecco and Liane proposed a rectanguloid design consisting of a 4 by 4 by 3 arrangement of the cubes. Whereas Ecco established the original adjacency guarantees, Alan Dragoo's solution was able to establish 15 additional ones. His design involved five levels: The first was 4 by 3 but with three missing blocks; the second 4 by 3; the third 4 by 3; the fourth 3 by 3 with an extra block; and the fifth 4 by 2 with many gaps (five blocks in total). Clever guy!

Here are the locations of Dragoo's cubes starting with the first floor:

	E	A
E	E	E
E	L	A
	L	

second floor:

C	B	A
N	J	A
E	I	A
E	H	A

third floor:

C	C	A
F	F	A
E	K	A
E	O	A

fourth floor:

	C	
E	D	D
E	G	G
E	M	M

and, finally, fifth floor:

	C
E	C
	C
	C

Inspiration and Offshoots

An architect once told me that technology made architecture ever more difficult by providing too much freedom—architects must contend with too many design choices. This puzzle removes the constraint of gravity, thus increasing freedom, but adds the constraint of connectedness. It is remarkable that mere mathematical exploration leads to better and better solutions.

8

The Tile Floor of Constantinople

Distinguished in his Armani tailored suit, Rino Banti showed us his badge: Servizio Speciale, the Italian Secret Service. "There is a little known secret about art in Italy," he explained. "All people have seen Michelangelo, Da Vinci, Tintoretto, but what you see is a tiny portion of what was produced. Looters from all over Europe have taken the patrimony of Italy for their own enjoyment. Not to mention local thieves. Sometimes the looting is destructive as in the case of the splendid Arabic-Venetian marble rectangle that is the subject of my visit.

"The Venetians, you see, instead of fighting in the Crusades, contented themselves with profiting from them. Their traders had friendly relations with the Moors, in fact, since they were partners in the spice trade. And quite a profitable trade it was. A kilo of spice bought in India for 5 ducats would fetch 100 times that price in Paris. The Venetians and Arabs found it easy to share these profits and many Venetian merchants fell in *amore* with Moorish design.

"The work of art we're seeking belonged to one of the great Doges of Venice, Enrico Dandolo, the blind conqueror of Constantinople in 1204. His role in the conquest was entirely exaggerated in fact.

The fighters of the fourth crusade conquered the city and turned it over to Venice in payment for the warships Venice had provided. The artwork is a geometric design in a rare turquoise marble. Dandolo couldn't see the turquoise, but it is said that he relished the touch of the lines carved in the marble.

"This floor, a 16 by 16 square, consists of smaller rectangles and triangles. These smaller pieces have been squirreled away in private collections all over the continent and some even in the United States. Most of the people who have them deny any knowledge of wrongdoing and we cannot, *francamente,* accuse them of lying. The theft occurred long before the pieces in question fell into the hands of their current owners.

"What's worse is that the owners protest *molto* about the prospect of giving up their works of art. Sometimes they have paid dearly for them. We think, however, that we can convince the courts that the correct pieces come from the stolen square provided we can prove that those pieces reconstruct the square and that no combination of the other turquoise pieces that we have located can.

"The trouble is that we don't know which are the correct pieces. At least some of the pieces we have located must belong to lesser works of art, because our mathematicians tell us they cannot form a square of the correct size. So, our problem is to figure out which pieces can possibly make up the square and how to form the square from them. Then we must show that no other combination of pieces can fit the square."

"Geometry!" Liane exclaimed. "I've been longing for a geometrical puzzle."

"So have I," Ecco said, smiling. "Signore Banti, please tell us the sizes of the pieces available as well as the size of the desired square."

"*Grazie molto,* Dr. Ecco," Banti said. "They told me you help only when you find the problem engaging. Please see this figure [Figure 1] for the sizes. . . . Oh, one last thing, we believe only four of the available rectangles are involved."

Cybernovice: Find a subset of these shapes that forms the square. Show how to form the square. There may be many copies of the triangles. Is

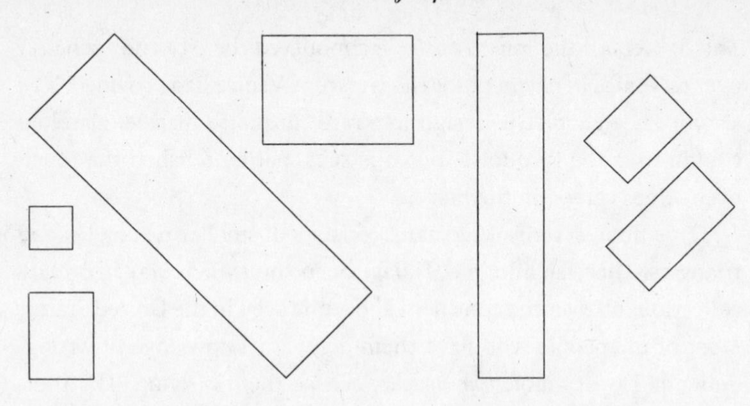

Figure 1. Suppose you are given a 16×16 square. You want to fill it with rectangles and triangles. Available rectangles measure 2×2, 4×3, 7×5, 16×3, $\sqrt{8} \times \sqrt{32}$, $\sqrt{18} \times \sqrt{8}$, $\sqrt{288} \times \sqrt{32}$. All triangles are right isosceles triangles. For every rectangle with an irrational side, there is a triangle whose hypotenuse matches that side. For every integral side of a rectangle, there is a triangle one of whose nonhypotenuse sides matches it. So, triangles are $2 \times 2 \times \sqrt{8}$, $4 \times 4 \times \sqrt{32}$, $3 \times 3 \times \sqrt{18}$, $7 \times 7 \times \sqrt{98}$, $5 \times 5 \times \sqrt{50}$, $16 \times 16 \times \sqrt{512}$, and $12 \times 12 \times \sqrt{288}$.

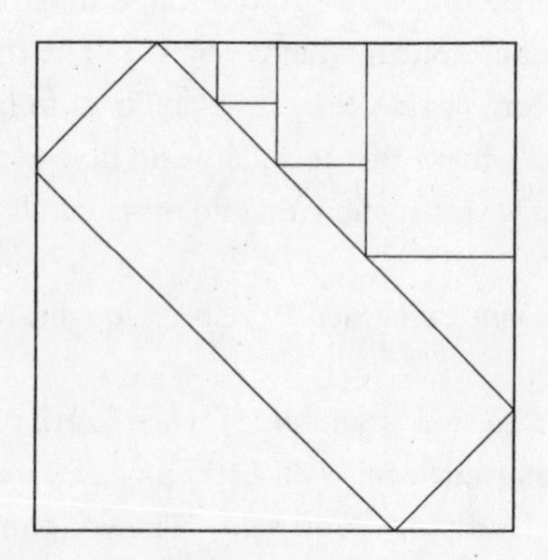

Figure 2. One possible solution to the layout problem using a total of 11 pieces of the following dimensions: (*rectangles*) 2×2, 4×3, 7×5, $\sqrt{288} \times \sqrt{32}$; (*right triangles*) $12 \times 12 \times \sqrt{288}$, $4 \times 4 \times \sqrt{32}$, $5 \times 5 \times \sqrt{50}$, $3 \times 3 \times \sqrt{18}$, $2 \times 2 \times \sqrt{8}$, $2 \times 2 \times \sqrt{8}$, $4 \times 4 \times \sqrt{32}$. The whole square is 16×16.

it the case that no other combination of shapes can possibly form the rectangle or must Banti find other evidence?

Solution

Figure 2 shows one possible solution. There are several more. See how many you can find. Banti will have to find some other proof.

Inspiration and Offshoots

Ask most scientists, engineers, or mathematicians whether they found tiles fascinating as a kid and they will say yes. This puzzle was born from this presumption of universality. The question here is how to reconstruct a kind of jigsaw. The solution is far from unique but could easily give the detective a head start. For example, suppose Banti found a document proving that only three rectangles were used.

9

Watchtowers in the Desert

Although he was two decades too old to fly fighter jets, the Air Force General Chris Warren had the confident smile and aviator glasses of a test pilot. "We've got a base to protect and we mean to do it," he said. "Sadly, even the best guards get bored, so we need several of them. Please look at this map."

He laid a piece of paper on a table and drew a big square.

"Our base is a square 76 meters on a side on a flat desert," he said. "Our most sensitive equipment—I can't tell you what it is—will be kept in a depot underground having a surface footprint of just a few square meters. We want the guards to alert our antiterrorist team if anyone approaches the depot. I pity the poor souls who get their attention.

"We have five guard towers to put up. From each guard tower, a guard can reliably see 18 meters. But each guard isn't looking everywhere all at once, of course, so we'll want several to cover some areas near the depot. We want the guards to be able to yell to one another, but not have such easy conversation that they forget

This adventure occurred in May 2001.

what they are doing. So, the towers should be at least 6 meters apart.

"We want at least 2,900 square meters to be visible to one or more guards, though different square meters can be visible to different guards. At least 775 should be visible to three guards or more. The original 2,900 will include these 775, of course. At least 100 should be visible to four or five guards."

"What does it mean for a meter to be visible, the whole meter, or just some portion?" Liane asked.

"For the purposes of this mission," the general answered, surprised by the 13 year old's question, "assume that each guard stands in the middle of a square meter [in a tower] and that seeing another square meter means seeing the middle of that square meter.

"Your assignment is to tell me where the towers should be, how many meters are visible, and by how many guards. If you can do much better than my specification, I'd be grateful."

Liane and Ecco found a design to satisfy the general's desires, but weren't sure they were doing as well as possible.

> *Advanced cybernovice: See whether you can find a design that has as many square meters as possible visible to one guard or more and still satisfies the constraints that at least 775 can be seen by three guards or more and at least 100 by 4 guards or more.*

Solutions

I start with Liane's solution and then improve on this with the readers' solutions. Starting where (0, 0) is the upper left-hand corner, here are the coordinates of the guard towers: (38, 36), (20, 22), (30, 49), (41, 30), (52, 24). So 2,911 square meters are seen by at least one guard; 1,242 by at least two guards; 777 by at least three guards; and 115 by at least four guards.

Figure 1 is the map of the area with numbers representing how many guards can see each point. The towers are indicated by asterisks.

Figure 1.

Now for some of the reader improvements. Dennis Yelle, Andrew Palfreyman, and Michael Birken were the first to give a solution in which 3,119 cells were visible to at least one guard. Birken reports: "I used a series of heuristics to try to speed up the search. For instance, if the total visible area was less than 2,900 or was less than the total visible area in the best solution found so far, then the guards tended to repel each other. Depending on other criteria, some tended to attract each other. The movements were governed by probability and the probabilities were updated by the heuristics."

Here is Yelle's solution: (18, 57), (35, 25), (36, 19), (57, 29), (29, 22), in which case 3,119 square meters were visible to one guard or more; 1,051 to two guards or more; 775 to three guards or more; 100 to four guards or more; and 0 to five guards.

Scott Williams used a slightly different heuristic to come to a similar answer: "The general placement is for 3 shacks to be placed fairly close to each other (to generate most of the 3+ meters). Place the 4th shack close enough to generate the 4+ meters (and the last few 3+ meters). Place the last shack completely isolated from the others to get the total meter count up." Tomas Rokicki found a property of his solutions: "All solutions covering 3,119 are equivalent through rotation, reflection, and translation of the two independent sections."

Inspiration and Offshoots

It is my experience in the commercial world that the more apparent security there is (e.g., the more guards), the less real security. Each guard thinks the other will do the job, so they all end up talking on the phone. So, I wondered how I would protect something really important. You might have a better idea.

10

The Lost Pillars of Menea

Apparently successful in spiriting her loot out of the Sudan, the famous archaeologist Natasha W was back in action, this time in Greece. "The famous temple at Menea was a center for games and prayer to Athena," the fit, tanned 30 year old said by way of introduction. "You, Dr. Ecco and Liane, the best puzzle minds in the world, should by rights help to reconstruct it, since Athena was the goddess of wisdom. Let me tell you about the building.

"From ancient texts, we know that the temple was built as a rectangle and the front pillars were exactly aligned with the back ones as you can see from my drawing. Put geometrically, the front and back were parallel lines, and a perpendicular line segment from every front pillar would hit a back pillar and vice versa. In line with the Greek love of regularity, the spacing between successive pillars (on both the front or the back) was designed to be as even as possible. In fact the ancient texts say the temple had seven pillars and speaks of the marvelous equality of their spacing. The lengths may have varied slightly because of the terrain or because of the local superstition that Zeus would be angry at a too perfect temple to

Athena—she was supposed to be his favorite, but then Zeus was fickle in his anger.

"Zeus's jealousy aside, the temple was viewed as a wonder in its time. History has not been kind to it, however. The beams crumbled in an earthquake and were much moved around by treasure-seekers, though we believe that no front beam was moved to the back or vice versa. Vandals and church building projects cannibalized the pillars.

"Our goal is to reconstruct the spacing of the pillars from the fragments of the beams. We have some hopes that this can be done for four reasons. First, in the original temple, each beam was a solid piece of stone from the center of one pillar to the center of the next one, except the end beams. Second, each beam broke into just a few fragments—altogether there are only 20 in front and 17 in back. Third, the weather in Menea is particularly dry so the beam fragments have hardly been eroded. Finally, the beams were cut from plentiful local stone, so nobody helped themselves to them."

"Can you fit the fragments together based on the shape of their faces?" Liane asked.

"We have tried, but the textures are too similar in all of them," Natasha replied. "Also the fragments are almost exactly rectangular. All we can do is give you lengths."

"OK, so let's try an example," Liane suggested. "Suppose there were three pillars [so two beams originally]. Suppose the front fragments [numbered 1 through 4] had lengths 1:2.1, 2:3.1, 3:2.1, and 4:1.8. That is, fragment 1 was 2.1 cubits long, fragment 2 was 3.1 cubits long, and so on. Suppose the back beam fragments had lengths 1:3.9, 2:0.3, 3:1.4, 4:2.2, and 5:1.3. Then we might form two groups in front: 1 and 3, giving a length of 4.2; and 2 and 4, giving a length of 4.9. The back would then be grouped as 1 and 2, again a length of 4.2; and 3, 4, and 5, again a length of 4.9."

"Exactly," Natasha exclaimed. "You are even smarter than you were 2 years ago! Now for the real problem. Here are the front fragments [labeled for your convenience] with their lengths:

1:2.4	11:4.4
2:1.0	12:3.6
3:2.5	13:4.6
4:0.4	14:1.2
5:3.0	15:3.7
6:3.7	16:4.0
7:0.4	17:1.1
8:1.2	18:0.7
9:4.2	19:3.3
10:2.4	20:0.1

"Here are the 17 back fragments in similar format:"

1:1.4	10:1.3
2:4.6	11:4.3
3:2.8	12:1.4
4:2.4	13:3.3
5:3.1	14:2.6
6:2.1	15:2.2
7:3.9	16:2.0
8:6.8	17:2.7
9:1.0	

1. Cybernovice: Can you find groupings of the beams in front and back, to account for seven pillars (including corners) in the front and in the back, whose spacings are all as close to equal as possible (minimum variance)?

"Excavation of the site uncovered a remarkable manuscript, preserved under the rubble on papyrus. This manuscript claimed that two pillars—in both the front and back, of course—were separated by more than 11 cubits and every pair by at least 4 cubits."

2. Cyberexpert: Can you find a minimum variance solution after adding this constraint?

Solutions

1. Here is one solution proposed by reader Michael Birken for the minimum variance, starting with the leftmost beams in the front

and back:

> [8:1.2, 3:2.5, 9:4.2], [12:1.4, 15:2.2, 11:4.3];
> length = 7.9.
> [19:3.3, 13:4.6], [13:3.3, 2:4.6]; length = 7.9.
> [14:1.2, 1:2.4, 11:4.4], [9:1.0, 16:2.0, 4:2.4, 14:2.6];
> length = 8.0.
> [2:1.0, 5:3.0, 16:4.0], [6:2.1, 3:2.8, 5:3.1]; length = 8.0.
> [18:0.7, 12:3.6, 15:3.7], [1:1.4, 17:2.7, 7:3.9];
> length = 8.0.
> [20:0.1, 4:0.4, 7:0.4, 17:1.1, 10:2.4, 6:3.7],
> [10:1.3, 8:6.8]; length = 8.1.

Birken was able to show that the minimum possible variance is 51/10,800 = 0.00472222222222... I quote his proof:

> The length of the assembled temple is 47.9 cubits (i.e., the sum of lengths of the fragments exclusively from the front or from the back). Since there are 6 beams in the assembled temple, the mean length of any beam is 47.9/6 = 7.98333333333333... Since the fragment lengths are all rounded to the nearest tenth of a cubit, the minimum beam length that does not exceed this mean is 7.9 cubits. By the same token, the maximum beam length that does not exceed this mean is 8.0 cubits. Ideally, we would like a solution containing lengths of [7.9, 8.0, 8.0, 8.0, 8.0, 8.0] since 7.9 + 8.0 + 8.0 + 8.0 + 8.0 + 8.0 = 47.9. Such a solution, if it existed, would have the minimum variance possible because any shuffling of a tenth of a cubit would either result in a solution with the same variance or a solution that had beam lengths that deviated even more from the mean of 47.9/6. Unfortunately, such a solution does not exist. One of the fragments from the back of the assembled temple is 6.8 cubits long. If it originated from a beam of length 8.0, then we must be able to assemble some of the remaining back fragments into a length of 8.0 – 6.8 = 1.2 cubits. Since the two shortest fragments from the back are 1.0 and 1.3 cubits, there is no way to assemble a beam of length 8.0 using the 6.8 cubit fragment. Similarly, there is no way to assemble a beam of

length 7.9 using the 6.8 cubit fragment since 7.9 – 6.8 = 1.1 cubits. Since the minimum length we can shuffle is a tenth of a cubit and the borrowed tenth of a fragment cannot be donated to the 7.9 cubit beam because that would result in a solution of the same form, the next possible solution with minimum variance is either of the form [7.8, 8.0, 8.0, 8.0, 8.0, 8.1] or of the form [7.9, 7.9, 8.0, 8.0, 8.0, 8.1]. The former has a variance of approximately 0.008056 and the latter has a variance of approximately 0.004722. So, if the latter solution exists, then it must have the minimum variance possible. And over 100 of these exist.

2. Reader Thomas Rokicki showed that with the added constraint that each beam is at least 4 cubits long and there is at least one beam over 11 cubits, the variance can be under 2.

[1:2.4, 6:3.7, 8:1.2], [1:1.4, 3:2.8, 5:3.1]; length = 7.3.
[2:1.0, 11:4.4, 17:1.1, 18:0.7, 20:0.1], [2:4.6, 17:2.7]; length = 7.3.
[3:2.5, 14:1.2, 15:3.7], [4:2.4, 9:1.0, 12:1.4, 14:2.6]; length = 7.4.
[7:0.4, 5:3.0, 16:4.0], [7:3.9, 10:1.3, 15:2.2]; length = 7.4.
[4:0.4, 10:2.4, 13:4.6], [6:2.1, 13:3.3, 16:2]; length = 7.4.
[9:4.2, 12:3.6, 19:3.3], [8:6.8, 11:4.3]; length = 11.1.

Inspiration and Offshoots

Have you ever been to a museum of natural history and wondered how the scientists reconstruct dinosaurs so precisely? Like paleontology, archaeology seems to be part collection and part construction. Constructions, like hypotheses, may have to be revised based on new facts. How would you resolve this one, if a document told you that both the front and back had a beam length of exactly 10 cubits, but the others were as even as possible in length?

Part III

Routing and Networks

11

Hovertanks in Battle

He claimed to be a military man, but Captain Solo didn't seem to have the right bearing. He slumped in a chair with his feet propped one atop the other. His uniform was clean, but was badly wrinkled. His hair was unkempt. In brief, he looked like one of my colleagues.

His smartly dressed assistant, Lieutenant Hood, presented the problem to Ecco. "Here is the layout of the valley, sir," he said. "Our hovertanks will be approaching from the eastern pass. When our adversaries detect us, they will approach from the western pass.

"Our job is to destroy the underground factories in the valley. The factories themselves have no defenses except their 1-meter-thick walls, but we must quickly dominate the ground above. That means we must rapidly reach a situation in which we can destroy any new vehicle that comes into the valley."

"Why can't you just cross over with your first hovertank and block their pass?" Liane asked.

"They have field guns and rocket-propelled explosives inside bunkers overlooking the western pass," the Lieutenant responded. "We have, however, located 25 hills that are good vantage points for

This adventure occurred in June 1998.

our hovertanks. Occupying any 10 of them will allow us to destroy the underground factories. We just want to be safe on them while keeping our adversaries to under 5."

"So, drive to the first 10 hills you can see," Ecco said with some impatience.

Solo spoke up. "We wouldn't be here if it were that simple, Dr. Ecco. You must think of this as a two-person game with alternating moves. We place a vehicle then they place one. Then we place one. Then they do. And so on. We just want to be sure that after they have placed 5, they can't place any more, whereas we can place at least 10 altogether."

"Can you get to any hill equally fast?" Liane asked.

Solo smiled. "I designed every hovertank to fly at 80 miles an hour and to stop on a dime."

"Unfortunately, our adversaries have access to the same technology," added the lieutenant. "Our biggest advantage is that we will be in the valley first. In fact, we hope that we will be able to place at least two and possibly three hovertanks on hills in the valley before they place any of theirs. After that we will alternate. Remember, we want to prevent them from placing more than 5 of their hovertanks and we want to be able to place at least 10 of ours. Also, our 10 hovertanks must be able to fire upon all unoccupied hills. Finally, their 5 hovertanks must not be able to fire upon our 10."

"Can you tell me which hills can fire upon which others?" Ecco asked.

"Yes," the lieutenant responded, handing Ecco the information shown in Table 1. "We've numbered the hills from 0 to 24. I hope the table is clear. For example, hill 0 can fire upon hills 1, 2, 3, 7, 8, 9, 13, 17, 18, 19, 20, 22, and 24. Note that firing is usually but not always symmetric. For example, 0 can fire upon 2 but the reverse does not hold. Finally, assume that a hovertank on any hill *h* can prevent an opposing hovertank from occupying *h*."

Ecco and Liane studied the table for a while. Suddenly, Liane giggled and said, "Two sets of queens problems." She then sketched

Table 1

FIRE FROM	FIRE UPON
0	1, 2, 3, 7, 8, 9, 13, 17, 18, 19, 20, 22, 24,
1	0, 2, 3, 9, 10, 13, 14, 15, 16, 17, 20, 21, 24
2	3, 5, 9, 11, 12, 13, 16, 22, 23
3	2, 5, 7, 9, 11, 12, 13, 16, 22
4	5, 7, 9, 11, 12, 13, 16, 22, 23
5	2, 3, 4, 7, 9, 11, 12, 23
6	1, 2, 3, 7, 8, 9, 13, 14, 15, 16, 18, 20, 21
7	3, 4, 5, 9, 12, 19, 20, 22, 23
8	0, 6, 10, 14, 15, 17, 18, 19, 24
9	2, 3, 4, 5, 7, 11, 19, 20, 22
10	1, 2, 3, 7, 8, 9, 13, 15, 16, 20, 21, 22, 24
11	2, 3, 4, 5, 9, 19, 20, 22, 23
12	2, 3, 4, 5, 7, 8, 17, 18, 23
13	0, 1, 2, 3, 6, 7, 9, 10, 16, 18, 21, 22, 24
14	1, 6, 8, 15, 16, 17, 18, 19, 21, 22, 24
15	1, 2, 3, 6, 8, 10, 14, 16, 18, 19, 20, 22
16	1, 2, 3, 4, 6, 7, 9, 10, 13, 14, 15, 17, 19, 22
17	0, 1, 8, 14, 16, 18, 20, 21, 24
18	0, 6, 8, 13, 14, 15, 17, 21, 22
19	0, 8, 14, 15, 16, 20, 21, 22, 24
20	0, 1, 2, 3, 6, 9, 10, 15, 17, 19, 22, 24
21	1, 6, 10, 13, 14, 17, 18, 19, 24
22	0, 10, 13, 14, 15, 16, 18, 19, 20
23	2, 4, 5, 7, 11, 12, 19, 20, 22
24	0, 1, 8, 10, 13, 14, 15, 17, 19, 20, 21

something on paper that looked like two squares and showed it to Ecco, who nodded.

Ecco then turned to us and stifled a yawn. "Gentlemen, let me take a 3-, 4-, or 5-minute nap and then I'll get back to you."

When Ecco returned, he handed Captain Solo a piece of paper. "If you can place 3 hovertanks before your adversaries can place any, then take these 3 and you can prevent the adversary from occupying any hills, because your hovertanks will be able to fire upon every remaining hill with deadly accuracy and speed. If you can place only 2 before they can place any, then choose these 2. You will be able to keep the adversaries to under 5, and occupy 10 yourself, and be able to fire upon all remaining hills. If you can place only one before they can place any, then I don't think you can achieve your goals, but I'm not sure and, well, you haven't asked."

1. Cybernovice: Try to show Solo how he can dominate the valley entirely by occupying three hills before his adversary can occupy any. Cyberexperts can show how many such solutions exist.

2. Cybernovice: Try to show Solo how he can achieve the conditions of the problem by occupying two hills before his adversary can occupy any (and alternating thereafter).

3. Cybernovice: Can Solo achieve his mission if he can occupy only one hill at the beginning?

Solutions

1. When Solo can take three initially, his hovertanks should occupy 24, 14, and 5. Those hovertanks can then fire upon everything

else, preventing the adversary from taking any hills. Several readers showed that if Solo can occupy three hills to begin with, there are 40 such sets that can fire upon all other hills.

2. When Solo can take two hills initially, he should occupy 14 and 5. The adversary can then occupy only 10, 13, 20, or 0. If they occupy 0, Solo takes 10 and they can do no more. At that point, Solo can take 4, 6, 11, 12, 15, 21, and 23, giving him 10 in all. Similar reasoning applies if the adversary takes 10, 13, or 20.

Why the giggling and the nap? The basic configuration of the hills appears to be built from two chessboards, one 4 by 4 and one 3 by 3. Fire-upon relationships are then constructed like queen moves. This doesn't hold completely, but close enough to lead to solutions.

3. Finally, when Solo can occupy only one hill to begin with, he cannot accomplish his mission. Here is the proof following a combination of the formulations of Ralph Nebiker, Michael S. VanVertloo, and Christian Tanguy.

Define a safe hill as one that can fire upon all hills that can fire upon it. Both 0 and 16 are safe and the adversary will be able to capture at least one of them after Solo's first move. In either case, the adversary will later be able to capture another hill that will give the adversary control over 18 hills. This is enough to prevent Solo from accomplishing his mission.

Inspiration and Offshoots

I've never been in the military. When I was of soldiering age, I helped design mainframes for IBM. For me, the IBM choice was vastly superior. Still, I've long been fascinated by the military task, where speed, stealth, and surprise are often as important as sheer force. Sun Tzu demonstrated this idea several thousand years ago in the telegraphic classic *The Art of War*, a book worth many readings.

12

Lasers on the Utopia Basin

"NASA is preparing missions for 100 years from now," Emma Skylight explained. The appropriately named scientist described herself as an optical propulsion physicist. She came to Ecco's apartment with an armful of detailed construction and laser designs. "Among the propulsion mechanisms we have designed for space colonies are laser shuttles. A laser shuttle propels a vehicle at high speed for short distances, a few tens of kilometers, provided the terrain is flat. The Martian terraces of the Utopia Basin enjoy the right topography. In our design, vehicles will be launched from one space station and 'caught' by braking lasers on the other side. By the very nature of this transport mechanism, the air corridor that is taken must be linear. Further, for safety's sake, we don't want air corridors to intersect. Collisions would almost surely be fatal."

"Let me see if I understand," Liane said. "Between two stations, there can be only one air corridor. Between three stations there can be the three air corridors of a triangle. Among four stations in a square there can be corridors corresponding to the four sides of a square, plus one diagonal. The other diagonal is disallowed because it would cross the first." She paused for a moment, then added, "That would hold for any convex quadrilateral."

"Right," said Skylight, staring incredulously at the 11 year old. "At the same time, we want there to be as many air corridors as possible among the stations so that no station will be isolated. Let me tell you about the stations. Each one will have a ground area of 200 meters by 200 meters and a height of 100 meters. The laser shuttles will fly to and from the roof. No shuttle may overfly another station. The stations will be placed at points in a large grid. The grid points are separated by 10 kilometers in the *x* and *y* coordinates.

"We have 20 points. They are not laid out very symmetrically I admit, but they follow the contours of the most temperate Martian valley.

"What we seek from you is a way to lay out the air corridors, so there are as many as possible and no station has fewer than three air corridors connecting it to other stations. Further no trip should require more than three shuttle hops—at most two intermediate stops."

1. Cybernovice: Would you like to give it a try? The air corridors should be straight edges in a graph while meeting the constraints, including the absence of crossing edges. (See Figure 1.)

2. Cybernovice: Can you show that no more than 45 edges are possible?

Solutions

1. See Figure 2, proposed by Doug Mewhort.
2. Burghart Hoffrichter proposed a nice proof to show that no more than 45 are possible: "By counting the number of horizontal and vertical edges in each rectangle and one diagonal per enclosed rectangle and the sides of the triangles, one sees that 45 is the maximum number of non-intersecting edges."

Other Musings

Other questions are possible as suggested by Denis Birnie of the very terrestrial Transportation and Traffic Systems Ltd in Christchurch,

Figure 1. Design routes for the laser shuttles using this layout of space stations.

New Zealand: (a) energy—minimize the number of lasers per station or in the network; (b) hub traffic—minimize the number of routes per station; (c) network load—minimize the number of routes per link. These could be applied to compare different 45-station solutions.

He also pointed out that safety is the main concern, so another goal might be to find "a 4-hop-to-anywhere network such that every journey can be completed by at least 2 routes with only the start and end stations in common."

An interesting conjecture proposed by Kees Rijnierse is that putting the stations slightly off the grid might make even better solutions possible.

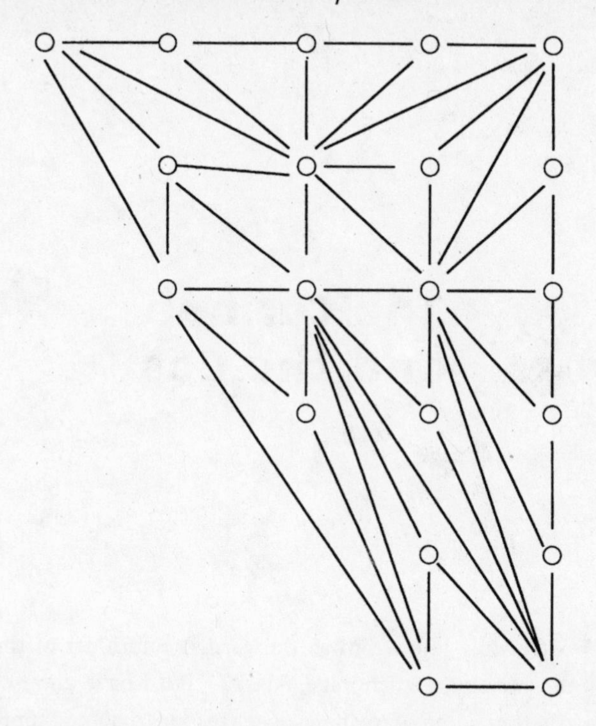

Figure 2. Here are 45 routes for laser shuttles that don't cross (graph is planar), give a maximum trip length of four, and such that each station has at least three edges.

Inspiration and Offshoots

I've long been fascinated with planar graphs (graphs whose edges never cross) of limited diameter. This graph is planar because the routes of the laser shuttles may not cross, as I think would be prudent in real life. In this case, a constraint beyond planarity is that the layout is fixed. It would be interesting to discover what improvements would be possible if one were allowed to move the stations just a little, as Kees suggests.

13

The Caribou and the Gas

Most people knew him as the garden columnist of the daily newspaper in Anchorage, Alaska. "But I have another identity that some of my newspaper readers may not approve of: I build gas pipelines," Alan Lionfall said. "Currently, when gas is discovered on the Alaskan north slope, it must be pumped back into the earth. I'm working on getting it into homes and power plants. Unlike oil, gas travels better underground at supercooled temperatures. The fewer the tunnels the better, especially in some of the areas precious to wildlife.

"Right now, I have a bit of a mess on my hand. You see, before I got involved, contractors had started laying some pipe through a caribou migration area. The pipes do not connect the right areas and some are probably unnecessary. Those contractors are so secretive that I don't even have a map of the area. All I have are these 'leg descriptions.' I finally understand what they mean.

"The positions are described by the letters A through M. We want to send gas from B to E—that's what the previous contractors didn't understand. Each leg is a one-way pipe from one letter position to another. Here are the pipes already laid:

A to F	G to J	K to C
A to H	G to L	K to D
B to M	H to C	K to H
D to L	H to I	K to J
F to E	J to B	K to M
F to G	J to D	L to C
G to A	J to M	L to I

"So, a leg is an edge in a graph," 12-year-old Liane said to Ecco, briefly stopping her strumming of the guitar. "From what I see there is no path from B to E."

Lionfall looked at Liane with a smile. "So, it's true she helps you solve the problems people ask of you, isn't it?" he asked Ecco.

"If it weren't for child labor laws, she might put me out of business," Ecco said in a deadpan.

"In any case," Lionfall continued, "there is no path from B to E as the young guitarist says. We have a bunch of routes we could potentially use without disturbing the environment and that each cost less than $1 million. Naturally, we'd like to build as few as possible. I list the possible ones below. Which ones would you suggest we use?

K to J	G to C	E to D
M to I	K to L	E to B
J to B	F to D	A to C
E to M	A to B	A to H
H to L	H to K	G to D
J to K	E to F	C to J
J to I	G to H	J to L
I to D	H to G	H to M
E to A	E to H	E to L
H to C	L to M	E to J
I to M	I to B	G to J
D to I	H to J	

Cybernovice: Dr. Ecco and Liane were able to get away with building only five additional routes. Can you match them or do better?

Solution

Here are the five additional routes Ecco and Liane came up with:

M to I
I to D
C to J
J to K
H to G

In conjunction with previous routes, these complete the following path from B to E: B to M, M to I, I to D, D to L, L to C, C to J, J to K, K to H, H to G, G to A, A to F, and F to E.

Inspiration and Offshoots

There is a famous topological theorem that holds that if you put your left hand on the entrance to the wall as you enter a maze and continue to walk while touching the wall with that hand, you will eventually find the outside exit. (This works for your right hand, too, but you might want to keep your right hand free to hold a flashlight.) The theorem is based on the observation that you will never touch the same portion of wall twice and there is only a finite amount of wall. The trouble is that this makes mazes mathematically uninteresting. To make them interesting again, I thought of the puzzle of creating a maze with no solution but giving the solver the right to pierce holes in interior walls in some number, say three, places. The question then—even to a solver who had a fragmentary map of the maze written down—was, "Which walls and where?" I still think this is a valid idea and would make a wonderful book or computer game, but I am too poor a maze artist to carry this out. The tundra problem is meant to be a topological cousin of this idea where instead of piercing holes in walls, one can add edges to connect a source (entrance) to a destination (exit).

14
Escape Art

The warden was a gray-haired man with smile wrinkles. The friendly pastor look and manner failed to match my image of the guardian of violent criminals. His companion, by contrast, had a hard face locked in a permanent scowl. He introduced himself only as "from the FBI" but volunteered neither name nor badge.

In any case, the warden began the conversation: "Cliff River is our state-of-the-art federal prison for high-security prisoners. We use robots for guards to eliminate any chance of corruption. People back up the robot guards with video monitors. There is a river on the south side that is impassable to boats and would drown the strongest swimmer. There is a cliff on the north side with a wall in front of it. The east and west sides have two rows of rolled barbed wire. Yet prisoner Doe apparently slipped out of his cell, traversed at least 19 others, and jumped off the cliff on the north side to safety with a parachute."

"Worthy of James Bond," 12-year-old Liane volunteered, one leg over the arm chair.

"Right, young lady, except this is a bad guy," the warden said. "He's a notorious hit man who has served as an executioner for the drug lords."

"The thing is the prison is quite secure," the warden said, shooting a glance at the FBI man whose mouth had settled into a sneer. "Let me tell you more about its design. It is organized as a one-story matrix: 20 north-south rows and 35 east-west columns. The cells are represented by rectangles and the hallways by line segments. Each prisoner is in a cell block by himself. The cell blocks themselves are concrete rectangular blocks so the prisoner cannot see outside into the hallway. He has a bed, a toilet, and a bare lightbulb. The door is on the south side. Books are permitted only on good behavior."

"The constitution allows this?" Ecco asked, his cheeks reddening. The FBI man looked at Ecco with narrowed eyes.

The warden seemed to ignore the comment and went on: "The five robot guards are in constant patrol to check that no cell block door has been opened. They start at midnight, one at the west side of rows 4, 12, and 20, and one at the east side of rows 8 and 16. The prisoner was at cell (10, 1) to start with. So, he was on the south side and had to get to the north side without being seen by the robots.

"Now for the timing. The robots proceed across (the west ones go east, the east ones go west). Then, when they cross completely, they go down one row, then they cross back, and so on for four rows at which point they return to their starting point. So, the southernmost robot starts at (0, 4) [see the bold segment], crosses to (35, 4), then goes to (35, 3), then crosses to (0, 3), then to (0, 2) and crosses to (35, 2), then to (35, 1) and crosses to (0, 1), and then returns to (0, 4). Since it checks each door, altogether it is checking 4×35 or 140 doors. The robot moves smoothly, taking 10 seconds per door and 40 seconds to do all the necessary travelling between rows, so it completes the whole task in 24 minutes." [Figure 1 shows what happens after 1 minute. Each robot has visited six doors corresponding to the line segments crossed.]

"While the robot moves east or west, it can sense anything that moves down the hallway and will shoot it, that's why we can't have more than one robot in a hallway at the same time. It cannot see backwards or sideways, though. The escaped prisoner moved faster:

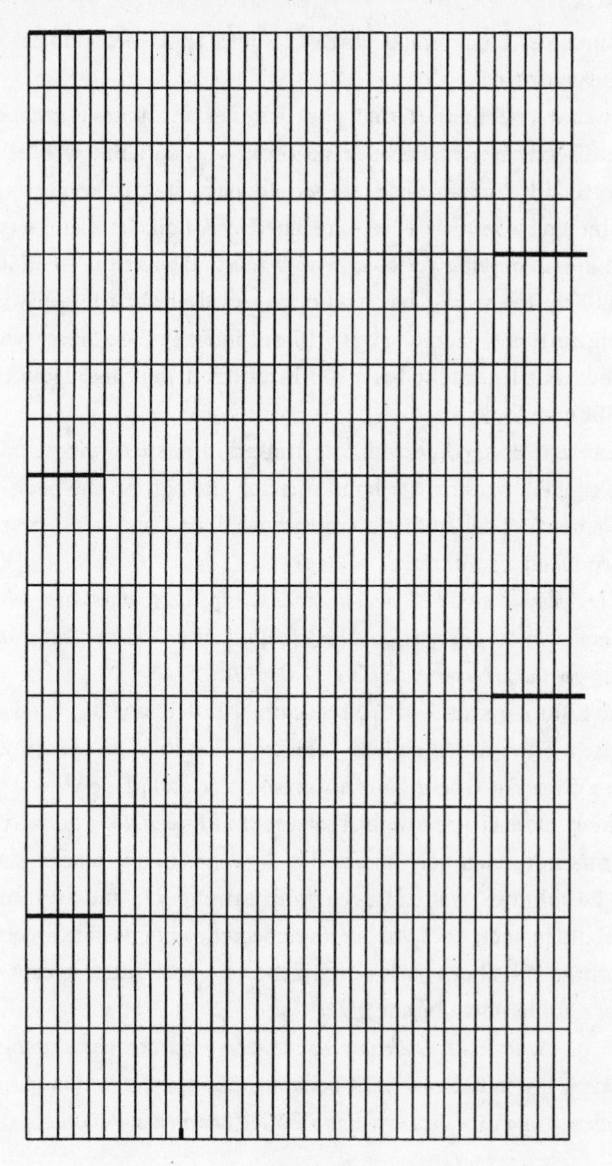

Figure 1.

5 seconds per block going north or south and 2 seconds per block going east or west.

"At first, we figured that the prisoner must have escaped by going directly north, because he broke through the wall at position (10, 20). In that case, he would have had to wait until the west-moving robot at row 4 reached position 10 (100 seconds) and the east-moving robot at row 8 reached position 10 (250 seconds). Then he would have been able to dash for the top without waiting further, reaching it at 310 seconds. But we have evidence to show that he reached the top in under 4 minutes. How is this possible?"

Liane and Ecco conferred. They talked in hushed voices and then Liane giggled, passing her hand through the hair on the back of her uncle's head. "That's what he must have done," she said. Soon after, she produced a solution.

1. Cybernovice: Would you like to try to find the smallest time following midnight the prisoner could have reached the north side? Liane was able to get the prisoner out before 12:03:20.

The FBI man snatched the solution, grunted, and stormed out of the room. The warden smiled, "He has a reason to be unhappy. He's the one who put Doe in jail in the first place and Doe vowed to get him back. Doe will be caught I'm sure, but I need your help to make sure this never happens again. I still want all cell blocks checked every 24 minutes, but can I reorganize the routes and/or timing of the robots in such a way that escape becomes impossible? Note that two robots will shoot each other if they see each other except when they are going north or south."

2. *Cybernovice: Is there any way to reorganize the robots so they visit every prison door every 24 minutes while completely eliminating the chance of escape and avoiding shooting one another?*

Solutions

1. Patrick R. Schonfeld described the fastest scheme at 3 minutes and 16 seconds. The method would go faster if the prisoner could

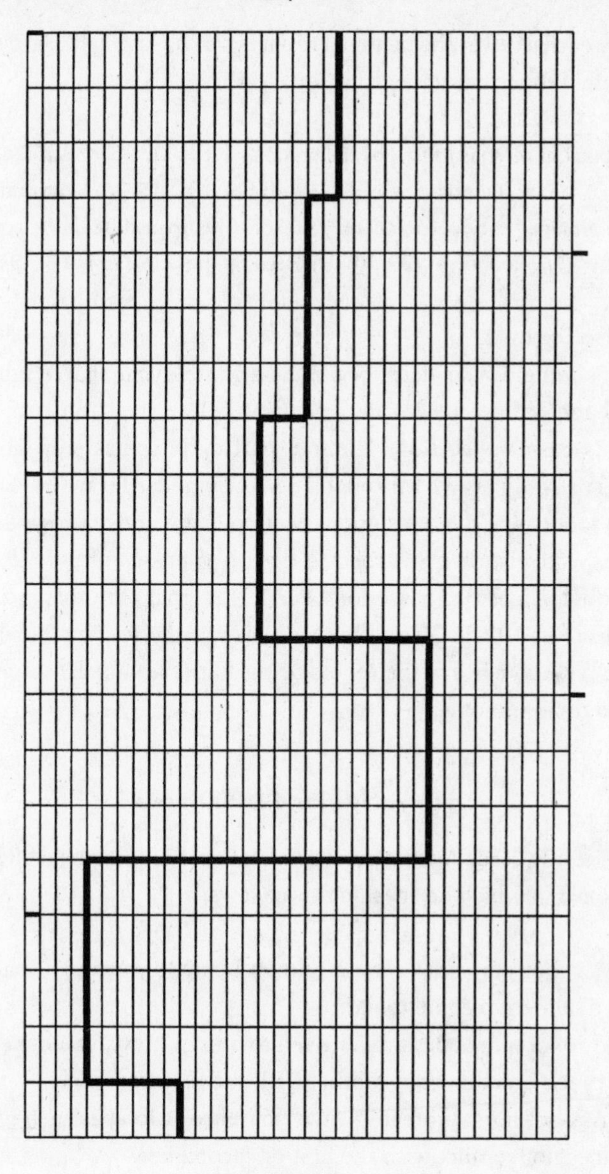

Figure 2.

safely reach a cell at the instant a robot is leaving the cell. Here's how Schonfeld describes it:

> Robot 1 at (4, 4) is passed by the prisoner at $t = 31$. Robot 2 at (26, 8) is passed by the prisoner at $t = 99$. Robot 3 at (15, 12) is passed by the prisoner at $t = 145$. Robot 4 at (18, 16) is passed by the prisoner at $t = 175$. Robot 5 at (20, 20) is passed by the prisoner at $t = 195$.

See also Figure 2.

2. Schonfeld also described a 24-minute cycle that eliminates the chance of escape for all prisoners below the 20th row. Here was his basic suggestion: The four-part cycle consists of (i) three robots patrolling rows 1 through 18 in the standard way (but faster), while robots 4 and 5 are stationary at opposite ends of rows 19 and 20, respectively, then (ii) robot 4 does two passes of row 19, then (iii) robot 5 does two passes of row 20. Note that with this cycle, the cells in rows 19 and 20 are checked twice per cycle. The fourth part of the cycle consists of the robots being stationary for 10 seconds to complete 24 minutes.

Inspiration and Offshoots

In the 1940s, Isaac Asimov wrote the delightfully original collection *I, Robot* in which he specified three laws:

1. A robot may not injure a human being, or, through inaction, allow a human to come to harm.
2. A robot must obey orders given to him by human beings except where such orders would conflict with the first law.
3. A robot must protect its own existence as long as such protection does not conflict with the first or second law.

Asimov's idealism has evaporated from modern engineering, unfortunately. What we are left with is the sheer inflexibility of the technology. That's what made the escape possible.

15

Emergency Response

Yves Maison introduced himself as a disaster planner. He didn't say which city he was planning a disaster response for, but it couldn't have been one of the old ones of his native France, because the city was laid out in a grid. "You see, Dr. Ecco," he began, "we have a new kind of ambulance that can carry two beds. We have 10 such ambulances, 4 at the Austerlitz Hospital, 3 at the Pasteur Hospital, and 3 at the De Gaulle Hospital.

"We serve our population very well in normal circumstances. But we have not a plan for disasters. That is my job." With that introduction, Yves lit a cigarette and sat down. We were all amazed to see a medical person smoking, but even Liane let it pass.

"Dr. Maison," Ecco said putting a plastic cup near his guest in lieu of an ashtray, "please tell us more about your city."

"We are laid out as a 100 by 100 grid," Maison replied. "Each grid point is an intersection, so the city is divided in blocks. Our hospitals are at positions (45, 45) for Austerlitz, (50, 50) for Pasteur, and (55, 55) for De Gaulle.

This adventure occurred in February 2001.

"If there is a disaster, we want to save as many people as possible, but perhaps not all. We in France remember La Grande Guerre [World War I to the rest of Europe] where doctors would triage patients into 'require immediate surgery,' 'can travel back to hospital,' and 'will die no matter what.' We view this as reality in any disaster.

"It takes 1 minute for an ambulance to travel one block. To rescue a person takes the time to get to the victim, plus 3 minutes to get him into the ambulance, plus the time to get him to the hospital, plus 1 minute to get him out at the hospital and deliver him to the emergency room. Each ambulance fits two victims. European ambulances often perform emergency care, but these disasters are anticipated to require hospital care. Each victim has a certain amount of time, called his survival time, beyond which it is no longer worthwhile to get him to the hospital. We want to figure out whom to rescue and by whom.

"We have a sample problem for you and would like to hear how many people you could rescue in time to enable them to survive. In our sample problem, we characterize the 30 victims by their geographic positions and their survival time [shown in the table below]. So, for example, victim 0—the first one—is at position (50, 55) and has a survival time of 35 minutes. If we can't deliver victim 0 to a hospital in 35 minutes, we don't attempt a rescue.

"Our question is how many people can we rescue, assuming that each ambulance starting at some hospital *h*, will pick up one or possibly two victims, if there is time for both, and return to hospital *h*? We are interested in both an answer and a method."

Cybernovice: Can you find six sites whose victims cannot possibly be rescued by any ambulance?

Ecco and Liane started discussing the method. I kept hearing words like "pruning" and "greedy," but didn't catch the details. Initially, Liane could save only 14 out of 30 victims, but eventually raised the number of saves to 18.

1. Cyberexpert: Would you like to try, given the initial distribution of ambulances, which is 4 at (45, 45), *3 at* (50, 50), *and 3 at* (55, 55)*?*

She explained her reasoning to Maison then added, "If you allow me to change the initial placement of ambulances, I can save 22 victims."

VICTIM	POSITION	SURVIVAL TIME (MIN.)	VICTIM	POSITION	SURVIVAL TIME (MIN.)
0	(50, 55)	35	15	(62, 51)	33
1	(48, 64)	42	16	(56, 56)	32
2	(49, 53)	32	17	(63, 61)	44
3	(53, 56)	39	18	(60, 51)	31
4	(53, 48)	31	19	(58, 53)	31
5	(51, 47)	28	20	(57, 72)	39
6	(52, 51)	33	21	(66, 60)	36
7	(52, 50)	32	22	(77, 56)	43
8	(52, 60)	42	23	(57, 62)	29
9	(47, 65)	42	24	(65, 65)	40
10	(57, 54)	31	25	(58, 69)	37
11	(69, 50)	39	26	(61, 56)	27
12	(57, 57)	34	27	(65, 57)	32
13	(56, 58)	34	28	(63, 70)	43
14	(64, 50)	34	29	(65, 56)	31

2. *Cyberexpert: How well can you do if you can choose the initial distribution of ambulances, but still under the rule that an ambulance leaves a hospital and then returns to the same hospital? Assume one can build an arbitrary number of ambulance bays at a hospital.*

Maison shook his head. "Unhappily, we have only a fixed number of ambulance bays at each hospital. That number matches the number of ambulances we have bought—four bays at Austerlitz, three at Pasteur, and three at De Gaulle. One thing you can do, however, is to stipulate that an ambulance can leave one hospital and deliver its patients to another one. Our emergency room can handle many patients. If more ambulances arrive at a hospital than there are bays, some may have to wait. It takes 1 minute to remove each victim from the ambulance [so, 2 minutes for two victims]."

I had to leave, so I never heard what Ecco thought about that one.

3. Cyberexpert: How about if ambulances from one hospital could deliver victims to another one, possibly increasing waiting time at the hospital bays? How many could you save then?

Solutions

Ecco and Liane's first approach in the case where an ambulance would leave a hospital, pick up victims, and then return to the same hospital was the following: Consider every pair of victims and every site and see which pairs could be picked up from which site within the survival times of both victims. Then the program would pick among these pairs provided the choice produced no conflicts. Use the remaining ambulances to pick up single victims.

1. Given the initial distribution of ambulances, Ecco and Liane were able to figure out how to pick up 14 victims. They used all three ambulances to pick up two victims from Pasteur and from De Gaulle, but none of those from Austerlitz. From Pasteur, remembering that counting starts at 0, they picked up victims 4 and 5 (at positions (53, 48) and (51, 47), respectively), as well as victims 6 and 8, and 7 and 10. From De Gaulle, they picked up 15 and 18, 16 and 29, and 19 and 26. From Austerlitz, two of the four ambulances each picked up one victim: 0 and 2. The other two ambulances did nothing.

But this is not the best solution. With the initial configuration of ambulances and the requirement that every ambulance had to return to its home site, Dennis Yelle and Andrew Palfreyman were the first to be able to save 18 victims using the following allocation of ambulances (the line "De Gaulle 1: 16, 10; 13, 12" means that the first ambulance from De Gaulle first picks up victims 16 and 10, returns to De Gaulle, then goes out and rescues 13 and 12):

Austerlitz 1: 0
Austerlitz 2: 2
Austerlitz 3: 4
Austerlitz 4: 5

Pasteur 1: 8, 6
Pasteur 2: 18, 7
Pasteur 3: 1
De Gaulle 1: 16, 10; 13, 12
De Gaulle 2: 26, 19; 3
De Gaulle 3: 29, 27

2. For the second problem in which ambulances can be based wherever they are most useful, Onno Waalewijn and Andrew Palfreyman suggested a way to save 22:

Pasteur 1: 2, 0; 6
Pasteur 2: 5, 4; 7
De Gaulle 1: 16, 3; 13, 12
De Gaulle 2: 8, 1
De Gaulle 3: 18, 15
De Gaulle 4: 23, 10
De Gaulle 5: 26, 19
De Gaulle 6: 29, 27
De Gaulle 7: 9
De Gaulle 8: 14

3. Dennis Yelle came up with the best solution—saving 21 victims—when each ambulance had to start at its home site but could drop off victims at any hospital provided it could find available bays. Here I'm using his notation in which h1 is Austerlitz, h2 is Pasteur, and h3 is De Gaulle. So, the second line means the ambulance starts at Austerlitz, rescues victims 2 and 0, transports them to De Gaulle, and then returns to rescue victim 3, who is also transported to De Gaulle.

h1 1 h2
h1 2,0 h3 3 h3
h1 5,4 h2 7 h2
h1 6,10 h3
h2 9 h2

h2 17 h3
h2 18,15 h3
h3 19,12 h3 13 h3
h3 26,16 h3 8 h3
h3 29,27 h3

Inspiration and Offshoots

Every time there is a disaster or war, health authorities must allocate resources. When resources are scarce, rescuing everyone is impossible. The general problem of designing a program to allocate ambulances in an emergency is NP-complete, meaning that finding the best solution requires examining all or a large fraction of possible schedules to find the best one. Good heuristics start with a greedy approach in which easy rescues are tried first in the hopes that there will be ample time to do multiple rescues. Clever readers often go beyond good, however.

16

The Police of Safetown

The newly elected mayor was young, clean-shaven, and round-faced. His slight chubbiness could be explained by the many apple pies and pork ribs he had had to eat during his election campaign. "The good people of Safetown want our fine city to live up to its name," he said. "As a candidate I promised to make that happen and I intend to keep that promise. The crime rate has gone up and the citizens are demanding that the police cruise through the streets with greater regularity, especially at night. The police in turn complain that with all the strange one-way streets, they can't figure out a way to cruise efficiently."

"So they must all drive?" 12-year-old Liane asked.

"Yes, we have fine one-way streets that I intend to give high-quality maintenance to—in fact I will raise the budget for asphalt-laying by 30 percent. . . ." He went on for a few more sentences, but I don't recall them.

"In any case," he continued, "police patrol their beats by car and we have three cruisers. We want to patrol every street as often as possible. It takes about 1 minute to drive down one block."

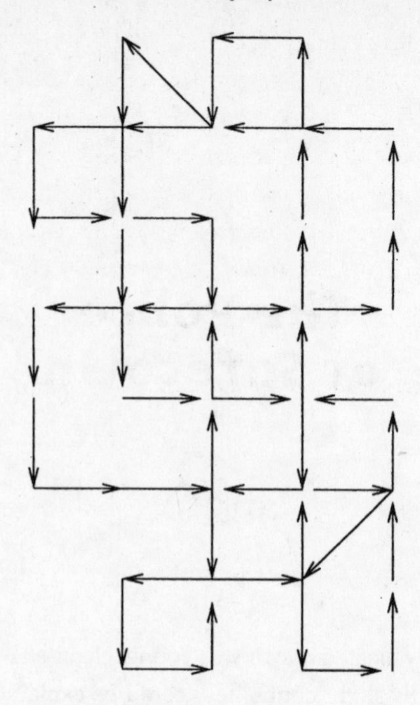

Figure 1. Road directions in Safetown.

"Well, let's see," said Liane looking at the map (see Figure 1). "Driving down each block every 16 or 17 minutes would be the greatest possible frequency, I think, given that you have 50 driving blocks. Shall we try for that?"

The mayor nodded and Ecco sat back as Liane started to draw.

Cybernovice: Can you arrange routes for the police so each car goes through its entire beat every 16 or 17 minutes?

The mayor studied Liane's solution. "And such a pretty girl, too," he said as he pumped her hand. "It would be my great pleasure to present you with the key to the city any time you come to visit. We could invite the high school marching band for the music and the Friends of the Police to make speeches. If we wait a year, I could use the event to announce my candidacy for Congress."

Eyebrows raised, Liane backed away slowly until she fell into a chair. She picked up her guitar and started to strum, slowly.

"So, I'll be calling you, young lady," the mayor said, pointing his finger at her. "And thank you, my fellow Americans." He left.

"Liane," Ecco said, after giving her a few moments to stare at the door, "suppose the mayor had five police cars. Can you arrange it so that the same police car drives down any given street at least once every 10 minutes? Different cars may drive on the same street, but we want each street to have its favorite car and that one drives down the street every 10 minutes. If not, what frequency can you guarantee?"

Cybernovice: Try for a 10-minute solution or show that none can exist. What is the best solution?

Solutions

1. Figure 2 shows the beats for three police cars taking 17 or fewer minutes each.

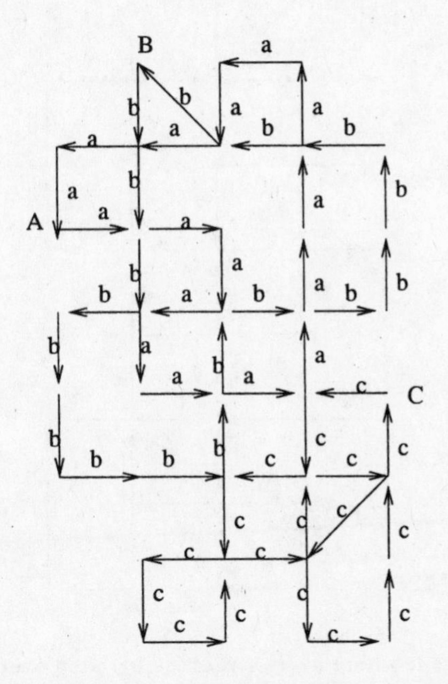

Figure 2. Solution with three police cars. No car traverses more than 17 streets. Each car (A, B, C) is represented by a letter.

2. Pike Enz was the first to show that every street could be visited every 13 minutes or less by the same car. Here was his solution:

Identifying vertices by x, y coordinates with the origin in the lower left corner,

(a) An 8-minute circuit: (1, 4), (0, 4), (0, 3), (0, 2), (1, 2), (2, 2), (2, 3), (2, 4), and back to (1, 4).

(b) An 8-minute circuit: (2, 2), (2, 1), (1, 1), (1, 0), (2, 0), (2, 1), (3, 1), (3, 2), and back to (2, 2).

(c) A 9-minute circuit: (3, 1), (3, 0), (4, 0), (4, 1), (4, 2), (4, 3), (3, 3), (3, 2), (4, 2), and back to (3, 1).

(d) A 12-minute circuit: (1, 3), (2, 3), (3, 3), (3, 4), (3, 5), (3, 6), (3, 7), (2, 7), (2, 6), (1, 6), (1, 5), (1, 4), and back to (1, 3).

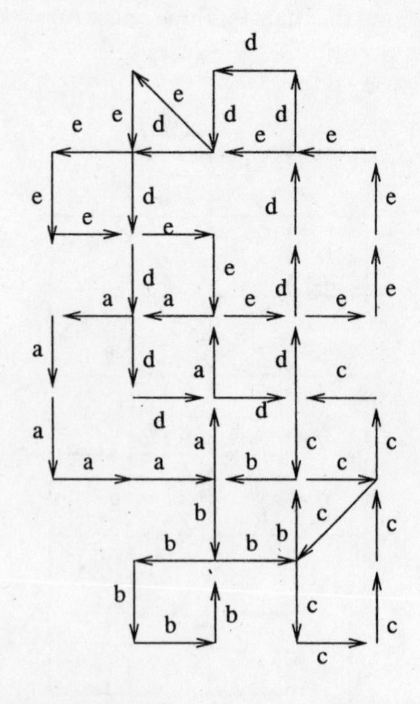

Figure 3. The case where every street is patrolled every 13 minutes or less by the same police car.

(e) A 13-minute circuit: (2, 4), (3, 4), (4, 4), (4, 5), (4, 6), (3, 6), (2, 6), (1, 7), (1, 6), (0, 6), (0, 5), (1, 5), (2, 5), and back to (2, 4).

See Figure 3.

Inspiration and Offshoots

Since Hammurabi in the West and Chin in the East, rulers have sent police—under a variety of names—to the streets to project authority. New York City, for example, puts its police on horseback, where they cut an imposing if vulnerable silhouette. Suburban towns prefer patrol cars that visit streets more or less frequently. As this puzzle shows, having all cars travel through all roads may be more efficient than giving each car a fixed beat of a portion of the town. A variant of this puzzle asks what happens if certain roads must be visited more often than others. How many cars would you need for this puzzle if, for example, all the streets whose y coordinate was above 3, that is, $(x, 3), (x, 4), \ldots, (x, 7)$ had to be visited every 10 minutes whereas the others could be visited as seldom as once every 20 minutes?

Part IV

Mathematical Geography

17

Of Politics and Fair Swedes

When I arrived at Ecco's house the first cool week of October, I was surprised but delighted to see Evangeline. She had spent the summer in the mathematics department at the University of Uppsala, Sweden. Now, she was back at Princeton pursuing what she called molecular logics. (She once tried to explain it as "logic without labels," but we were interrupted before I could hear more.)

"What surprised me most was that they really believe that philosophy is important at Uppsala," she was saying to Ecco. "They think that people who subscribe to the consensus-based socialism embodied in the Swedish political economy will invent a certain kind of mathematics."

"And do they?" Ecco asked.

"In subtle ways," Evangeline answered. "The pattern recognition techniques scientists develop there are based on Demster-Schaefer theory, which works by eliminating disagreements among different data sources."

"Maybe they're on to something," Ecco allowed.

"They asked me to work on a different kind of voting, Jake,"

Evangeline went on. "I think they wanted me to discuss it with you and it seems difficult enough that you might like to hear it."

Ecco nodded, "Please proceed."

"Götoldenborg is one of the oldest cities in Sweden," Evangeline said. "It is really a large village, with a population of about 170,000. That population has been shrinking of late, so the Swedish government has decided to reduce the number of districts from 66 to 28. The goal is to create districts with roughly 6,000 people in each, give or take 100 people. This is the problem they wanted your help with. Here is the map," Evangeline said. (See Figure 1.)

"The numbers are in a funny arrangement," Ecco said.

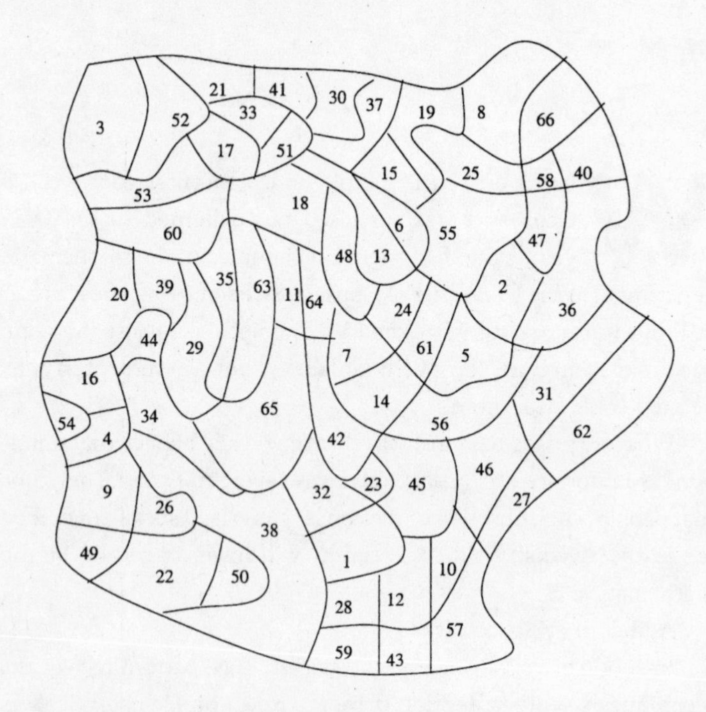

Figure 1. The original 66 districts. The population of each district is shown in the text. Your job is to reduce these 66 districts to 28 districts of roughly equal population (6,000 residents). It should be possible to walk from anywhere within a district to anywhere else within the district without going outside that district.

"I said that too," Evangeline replied, laughing. "They are numbered in the order in which each area became a town. Götoldenborg is really the conglomeration of all these small towns."

"What are the constraints on the solution?" Ecco asked.

"First, each district must be connected," Evangeline replied. "That is, it must be possible to walk from anywhere within a district to anywhere else within that district without passing through any other district, much as in the population exchange problem you mentioned to me a few months ago. Second, each district should have 6,000 people as I mentioned before. Third, you should never divide an existing district to create a new one. Here are the current populations:

DISTRICT	POPULATION	DISTRICT	POPULATION	DISTRICT	POPULATION
1	2,992	23	2,987	45	3,023
2	2,032	24	2,031	46	2,973
3	3,021	25	2,969	47	2,024
4	1,973	26	3,028	48	5,976
5	2,020	27	2,991	49	1,975
6	2,977	28	3,001	50	1,976
7	2,003	29	1,993	51	1,969
8	3,004	30	2,010	52	2,978
9	2,985	31	2,995	53	2,030
10	3,024	32	2,979	54	2,002
11	3,032	33	2,008	55	2,971
12	3,004	34	2,010	56	1,991
13	3,020	35	3,027	57	2,997
14	1,980	36	2,991	58	2,024
15	2,026	37	1,974	59	2,990
16	2,008	38	2,019	60	1,993
17	1,990	39	1,979	61	2,009
18	1,984	40	3,028	62	3,004
19	1,978	41	2,020	63	3,026
20	2,008	42	2,983	64	2,984
21	2,008	43	3,004	65	3,011
22	1,991	44	1,970	66	2,993

1. Cybernovice: Find a mapping from the original 66 districts to 28 districts such that each district has close to 6,000 inhabitants (within 100), each district is connected, and no existing district is divided up.

Though Ecco was able to do this, Evangeline returned a few weeks later with a more challenging problem. "It seems that 6,000 is too small," she said. "They want just 14 districts with as close to 12,000 residents as possible."

2. Cyberexpert: Please give this a try.

Solutions

1. You should be able to map original districts to new district numbers by staring at Figure 1. Figure 2 illustrates Ecco's solution:

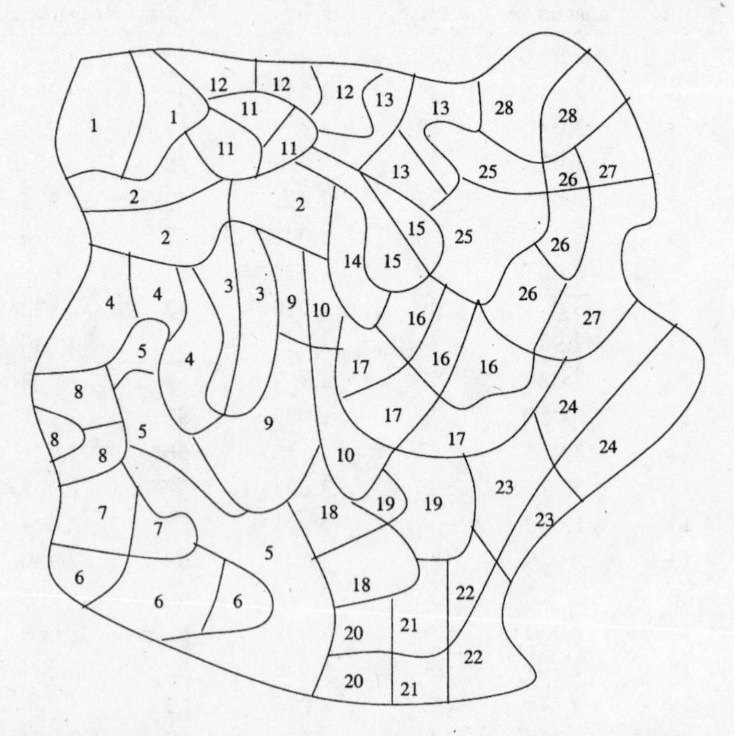

Figure 2. This figure shows Dr. Ecco's 28 district solution superimposed on the original map. Thus, all areas having the same number will belong to the same district, no matter what the original numbers were.

ORIGINAL DISTRICT NUMBER	NEW DISTRICT NUMBER	ORIGINAL DISTRICT NUMBER	NEW DISTRICT NUMBER	ORIGINAL DISTRICT NUMBER	NEW DISTRICT NUMBER
1	18	23	19	45	19
2	26	24	16	46	23
3	1	25	25	47	26
4	8	26	7	48	14
5	16	27	23	49	6
6	15	28	20	50	6
7	17	29	4	51	11
8	28	30	12	52	1
9	7	31	24	53	2
10	22	32	18	54	8
11	9	33	11	55	25
12	21	34	5	56	17
13	15	35	3	57	22
14	17	36	27	58	26
15	13	37	13	59	20
16	8	38	5	60	2
17	11	39	4	61	16
18	2	40	27	62	24
19	13	41	12	63	3
20	4	42	10	64	10
21	12	43	21	65	9
22	6	44	5	66	28

2. The best solutions to this problem came from readers Ralph Nebiker, Roger Alley, Friedrich von Solms, and Serguei Patchkovskii (who used exhaustive search). The populations are all remarkably close to 12,000. In fact, the maximum deviation from 12,000 these solutions produced was a surprisingly low 29. The numbers within parentheses are the original district numbers.

ORIGINAL DISTRICT NUMBERS	POPULATION	VARIANCE
11, 48, 64	11,992	8
8, 15, 19, 37, 66	11,975	25
2, 14, 36, 40, 56	12,022	22
10, 31, 46, 62	11,996	4
12, 27, 43, 57	11,996	4
5, 25, 47, 55, 58	12,008	8
7, 24, 32, 42, 61	12,005	5
1, 23, 28, 45	12,003	3
6, 13, 30, 41, 51	11,996	4
3, 17, 21, 33, 52	12,005	5
4, 9, 26, 34, 49	11,971	29
22, 38, 50, 59, 65	11,987	13
16, 20, 39, 53, 54, 60	12,020	20
18, 29, 35, 44, 63	12,000	0

Inspiration and Offshoots

Representative government is open to many abuses. For example, suppose the political system consists essentially of two parties: A and B. Party A is in the minority but can nevertheless achieve a representational majority by strategic redistricting: create many districts in which A has a bare majority and a few (preferably more populous) districts in which B holds enormous dominance. Beyond a majority, the B voters are useless to the B party. This puzzle is a population redistricting puzzle, because I think strategic redistricting is immoral, even though it is mathematically interesting.

A graph theoretical solution can form the basis for a computerized attack on the problem: The districts are nodes and two nodes are linked if they are adjacent. The question then is how to map nodes into supernodes comprising several adjacent districts and satisfying the constraints.

18

A Crisis in Middle Europe

With his silver hair, piercing eyes, and Oxbridgian prosody, William Pencil looked and sounded much the distinguished elder statesman. The furrows in his brow, though, revealed a sadness characteristic of someone who had seen too much suffering—or had felt it too deeply.

"The paradox of modern times is that communication and transport have become nearly instantaneous, enabling people to communicate and mingle as never before, yet ethnic tension becomes ever more brutish," he began in an oratical tone. "Two years ago, the Czech Republic and Slovakia both applied to the common market, an internationalist step. At the same time, the two peoples simmer with resentment toward one another to such an extent that friendships founder on the question of ethnicity. We see this pattern all over the world: Catalonia and Spain, Armenia and Azerbaijan, not to mention the Balkans, the Middle East, and south Asia. Statesmen can sometimes bring peace, but seldom for long. Sooner or later, I'm afraid, a population exchange takes place, often a bloody one. Witness the bloodshed after the Indian subcontinent became independent in 1948.

I want to reduce the pain of this process mathematically and I want your help.

"A new ethnic rivalry has arisen in middle Europe and we want to create a population exchange plan. We have divided the area into a 10 by 10 grid. Each cell has a homogeneous population. Remarkably, each cell also possesses roughly the same natural resources.

"We want to find the fewest exchanges required that give each group a continuous region. A region for group X is continuous if one can travel from any cell in region X to any other cell in region X without having to pass through the cell of another group."

Ecco stopped Pencil, "In your grid, do you count two cells to be adjacent if they border each other on the diagonal?"

"No, only if they border horizontally or vertically," Pencil answered.

I volunteered a clarification: "Ecco, by graph theoretically he means that each group should live in one connected component where each cell in the grid is a node and there is an edge between two nodes if they border each other horizontally or vertically," I said.

Pencil nodded and showed us a piece of paper. "Yes, the mathematicians have formulated the problem exactly that way. But that doesn't help bloody much. The question is how to achieve this state.

"The history of population exchanges is sad. In 1948, for example, mobs attacked the trains of Muslims leaving India and of Hindus leaving Pakistan. Also, the homes each side left for the future occupants were in horrible disrepair. Some were burnt to the ground. We want to manage this exchange much better: All exchanges will be pairwise. The heads of the exchanging cells will meet to assure one another that they will leave their villages in good shape. UN soldiers will try to prevent looting."

"If everyone is going to be so reasonable, then maybe they can just keep their homes and live in peace," Liane said.

Pencil let out a sigh. "I wish I could share your optimism, young lady," he said to 10-year-old Liane. "But the land suffered from riots last year and tensions are quite high."

"Very well, we will help you," Ecco said. "Please give us the current grid."

"Wait. I still don't understand the problem," Liane interrupted. "Could you give a small example?"

"Capital idea," said Pencil, smiling for once. "Please understand that we represent each group by a number 0, 1, or 2. Suppose we had a 6 by 5 grid:

2 1 2 1 0
1 0 1 1 2
1 2 1 1 0
1 0 1 0 0
2 0 1 2 1
0 0 0 0 0

"Here is a transformed state that meets our conditions:

2 2 2 2 2
1 1 1 1 2
1 1 1 1 0
1 0 1 0 0
1 0 1 0 0
0 0 0 0 0

"Notice that it is possible to go from any cell to any other of the same group by walking vertically or horizontally and remain within the same group at all times. How many exchanges would it take to achieve the transformed state?"

Cybernovice: Before reading on, please give this a try.

Liane thought for only a few minutes and then said, "Would four be too many?"

Cybernovice: Before reading on, see if you can figure out which four exchanges would arrive at the transformed state.

Liane continued, "Suppose (0, 0) is the top left corner and (5, 0) is the lower left corner. First, exchange the 2 at (4, 0) for the 1 at (0, 1), then (0, 4) for (4, 3), then (2, 1) for (0, 3), and finally (1, 1) for (4, 4)."

"Not bad," Pencil said. "Not at all bad. It took our best mathematicians a bit longer than it took you. Here now is our real problem. Please see what you can do. Any number of exchanges under 20 would be fine. We won't specify the transformed state. We just ask you to produce—what did you call them—a single connected component for each group. Please see how few exchanges you can use."

0 0 0 1 0 1 2 0 2 0
2 2 2 2 1 0 0 0 0 1
2 2 1 2 2 1 1 0 1 2
2 2 1 1 1 2 0 1 2 0
1 1 1 2 2 1 0 2 1 1
1 2 1 2 2 0 0 1 1 0
2 0 0 2 0 0 1 1 1 1
1 2 1 2 1 0 1 1 0 1
1 0 2 1 2 1 0 2 2 1
0 2 1 1 1 0 1 1 0 1

Ecco and Liane worked together and came up with an appropriate state after only 15 exchanges. "We really wish you could find a more imaginative solution to this problem, Mr. Pencil," Ecco said as he handed Pencil the solution. "Perhaps a national juggling team?"

Cyberexpert: Can you find 15 or fewer exchanges to achieve a state in which each group has a continuous region? The best solution I know of requires only 13 exchanges.

Solutions

Bruce Wilson and John Porter found the solution to the cybernovice question given in the text.

For the main puzzle, Ralph Nebiker offered a solution requiring only 13 exchanges, as shown below.

Final arrangement:

0 0 0 0 0 0 0 0 0 0
2 2 2 2 0 0 0 0 0 0
2 2 1 2 2 2 0 0 1 0
2 2 1 1 1 2 0 1 1 0
1 1 1 2 2 2 0 1 1 1
1 2 1 2 2 0 0 1 1 1
1 2 1 2 0 0 1 1 1 1
1 2 1 2 0 0 1 1 1 1
1 2 2 2 2 2 2 2 2 1
1 1 1 1 1 1 1 1 1 1

Exchanges:

(0, 3) for (6, 2)
(2, 6) for (9, 0)
(2, 5) for (6, 0)
(9, 1) for (4, 5)
(0, 8) for (8, 1)
(0, 5) for (9, 5)
(6, 1) for (0, 6)
(4, 7) for (8, 3)
(3, 8) for (8, 5)
(2, 9) for (8, 6)
(1, 4) for (9, 8)
(7, 4) for (7, 8)
(1, 9) for (5, 9)

Finally, Steve Worley suggested the following open problem: Suppose the only possible exchanges were between neighboring villages. What would the best answer be then?

Inspiration and Offshoots

As Gordon Allport observes in *The Nature of Prejudice*, "opposites attract" is mostly a myth. People like their own kind and, absent conscious effort, will feel emnity toward anyone different. Evidence abounds already in this millenium, virtually in every country. The only cause for hope is that persistent educational efforts can mitigate prejudice. In the meantime, how best can one disengage several groups at times of high tension? Ethnic cleansing followed by consolidation leaves thousands of people dead. It would be far better to organize an orderly consolidation in which areas are exchanged in a piecemeal fashion so each side would have an incentive to behave, thus minimizing loss of life and property. This puzzle was dedicated to this simple idea. Consolidation does not necessarily mean a tight cluster but at least connectedness. Tight clustering might follow later and introduces new mathematical constraints of convexity.

19

Enemies, History, and the Importance of Beachfront

He introduced himself only as Henry. "It is our policy, the policy of the government of the United States, that is, to eliminate all unnecessary wars on the planet," he said in that soft clear voice characteristic of aristocratic Connecticut Yankees. "On the other hand, we want to avoid expending ourselves in regional conflicts, as we did all too recently in the Balkans.

"Our goal, therefore, is to maintain the peace where it exists and establish it where it doesn't. Such is the role of the global superpower.

"In the difficult island of Aresia, there are 27 ethnicities. All are proud and warlike. They seem to obey the following rules of "friendship," which I use interchangeably with "alliance" in what follows:

1. Nonneutrality: I am either the friend or enemy of everyone else, but not both. If someone is my enemy, then I am his (her) enemy.
2. Transitive and reflexive: My friend's friend is my friend and I am my own friend.
3. Expediency: My enemy's enemy is my friend. (See Figure 1.)

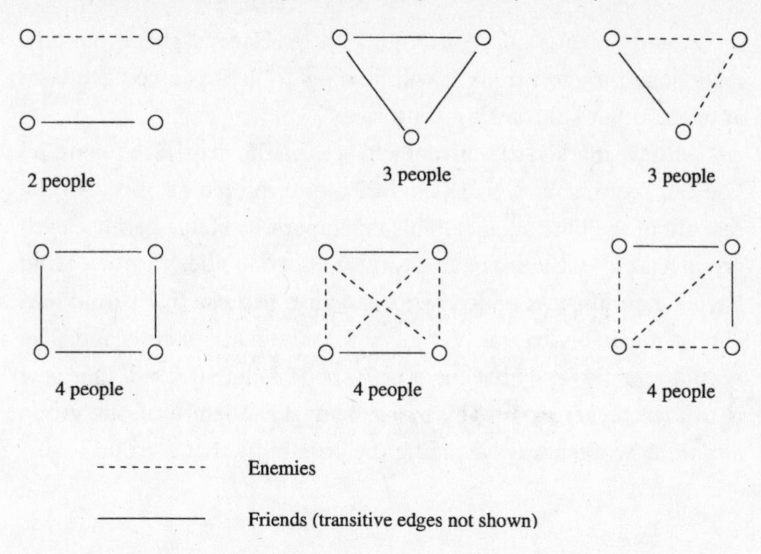

Figure 1. There are two configurations for two people (two enemies or two friends), two configurations for three people (all friends or two friends and an enemy), and three configurations for four people. Can you find the pattern?

"Our first question is: How many different allied groups can there be at any given time?"

> *Cybernovice: Before reading on, can you decide how many groups of friends there can be? Not their sizes, just how many groups?*

Ecco turned to Liane. She began speaking, thinking aloud: "Let's see. By 2, friends are transitive. By 1, groups of friends partition the space and friendship is symmetric. By 3, there can be only one or two partitions. If there were three, say A, B, and C, then they would all mutually be enemies, but then A and C would be friends, because they are both enemies of B. So, this cannot happen."

Henry looked at Liane, still an 11-year-old kid in her ponytail. "You are a national asset," he said appreciatively. "Is there any limitation on the number of these 27 groups that can be in each group?"

"No, all numbers seem to be possible," Liane said. "1 and 26, 2 and 25, . . . , 27 and 0."

"27 and 0 means all friends, right?" asked Henry. "That is our ultimate goal, though I think it will be tough. These people have been at one another's throats for centuries."

"We have marked the current alliances on the map. [See Figure 2.] You can see that 2, 8, 9, 15, and 20 are in the red alliance. All the rest are in the blue alliance. This is temporarily stable because each group has either the sea or an ally on at least one side. We must avoid having a group surrounded by its enemies, because that would lead almost inevitably to war. It's lucky that the place is an island. The seaside groups rarely have to worry about being attacked. Our goal is to make everyone an ally by changing the alliance of one group at a time, while always avoiding the possibility that a group is sur-

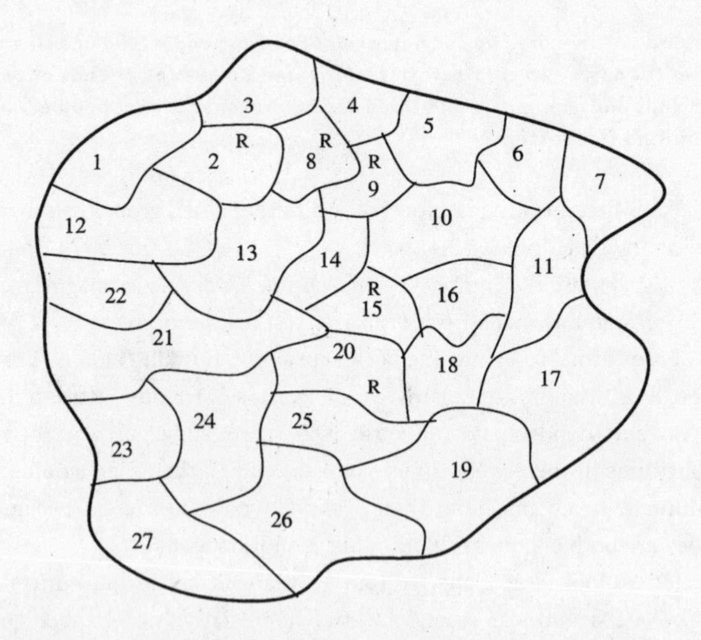

Figure 2. The alliances on the island of Aresia. All regions are in the blue alliance except those marked with R: 2, 8, 9, 15, and 20. Dr. Ecco must figure out how to bring peace by "encouraging" a few changes of alliance.

rounded by enemy groups. Since changing the alliance of a group is difficult and expensive, we want to do this as few times as possible.

"One of our mathematicians has suggested that achieving our goal at minimum cost should never require forcing an ethnicity to change alliances twice. What do you think of that conjecture?"

Ecco and Liane conferred for a while, then they responded.

1. Cybernovice: What do you think?

"How interesting," said Henry.

"Well, with that in mind what is the smallest number of alliance changes we must force to achieve our goal? Remember the caveat: Never leave a group surrounded by the hostile alliance on all sides."

2. Cybernovice: Dr. Ecco and Liane were able to do this by changing nine group alliances altogether (without leaving a group surrounded by its enemies). Can you do better?

"Nicely done," said Henry. "Some of our analysts think that the difficulty of changing a group's alliance is proportional to the group's population times the ratio of the source alliance's population to the target alliance's population. Here is a list of the populations on Aresia.

ETHNIC GROUP	POPULATION	ETHNIC GROUP	POPULATION	ETHNIC GROUP	POPULATION
1	56,727	10	21,708	19	93,559
2	77,232	11	61,758	20	53,964
3	105,454	12	50,790	21	78,913
4	50,660	13	169,466	22	155,511
5	65,707	14	82,659	23	92,888
6	24,692	15	95,675	24	90,596
7	82,943	16	63,247	25	64,713
8	61,871	17	104,643	26	105,709
9	17,781	18	144,026	27	101,417

Total reds initially: 306,523
Total blues initially: 1,867,786
Total: 2,174,309

"For example, if I want to change region 8 from red to blue, the difficulty is 61,871 × (306,523/1,867,786). The ratio reflects the fact that changing from the minority alliance to the majority alliance is much easier than vice versa, because citizens like to switch to the richer team. Given this notion of difficulty, which groups should we force to change alliances to guarantee peace [everyone allied] at minimum difficulty?"

3. Cybernovice: What strategy would you propose and what would the difficulty be? Dr. Ecco was able to find a strategy having a difficulty of less than 426,000. Remember that at no time should one group be surrounded on all sides by hostile groups.

Henry looked at the solution. "How curious! Groups on the periphery can carry out a much more consistent alliance policy. Maybe that's the secret of great empires."

Solutions

1 and 2. It is better to force two changes of alliance sometimes, for example, change 13 and 21 to red, then change 2, 8, 9 15, 20, 13, and 21 to blue. That requires only nine changes. Moreover, the ones who should change are those that are surrounded by the most potentially hostile neighbors. Is there a mathematical model of history here?

3. The following ordering presents the minimum difficulty of 309,803.55 for conversion: 10 to red, 6 to red, 20 to blue, 2 to blue, 8 to blue, 9 to blue, 15 to blue, 10 to blue, and 6 to blue. This was first pointed out by Jon Beal.

Inspiration and Offshoots

In the history of statecraft, alliances have shifted, sometimes due to war and sometimes due to changes in perceived interest. The fickleness of friendship should give pause to normal people who allow themselves to be whipped into a bloodlust by the warmongers.

Enemies, History, and the Importance of Beachfront

History shows how often one's enemies may later become one's friends. In modern times, this happens among powers that are roughly equal in strength. The biggest problem occurs when one side is vastly weaker than the other. These musings were the inspiration for this puzzle. Diplomats take note.

Part V

Scheduling

20

The Clock Gourds of the Amazon

"We have found a remarkable tribe in the jungles of Venezuela," the strongly built ethnobiologist Panta de Leon began in slightly accented English. de Leon had heard of Ecco from Natasha, the archaeologist of the Sudan. de Leon's motivation, we all decided later, was entirely pure. "The tribe has lived in near permanent isolation except for occasional, mostly unfriendly exchanges with outsiders. They are fine singers and percussion musicians. Their principal musical instrument is a fruit gourd called a *calabaza*. This gourd is so central to their life, in fact, that they call themselves the Calabazas when meeting [friendly] outsiders. Recently, I discovered that they also use the calabaza as a timepiece for the ritual they call—I'm translating roughly—'Prayers to the Clock God.'

"Let me explain. Theirs is one of the wettest, most mosquito-infested corners of the world. This has kept them safe. It also means that anything made of metal either rusts or is quickly impregnated with water.

"Many years ago, they found a group of explorers who had lost their way and then died after consuming a poisonous fruit. One of the explorers had a watch which the Calabazas brought back to the village. The watch worked for one moon, they say.

"The first day they had it, the tribal elders tried to figure out the meaning of the timepiece and they quickly concluded that it must be used for prayer. The explorers died because they hadn't prayed at the right time, they decided. Fascinated by the circular rhythm of the second hand, the elders directed the village artisan to carve out a hole in several gourds of the same size. They told the artisan that he should carve the gourds so that a full gourd would drain completely in 1 minute when the hole was opened. Unfortunately, the village artisan found himself unable to carve out a large enough hole without destroying the calabaza. Instead, he carved out two holes.

"Opening one hole in a full calabaza would cause the water to drain completely in 11 minutes. Opening both holes would cause it to drain in 5 minutes. These were exact times, to the second, in fact—a remarkable achievement considering that all his tools were handmade."

Liane perked up. "Why should two holes be more than twice as fast as one?" she asked.

"The vortex effect, I think," Ecco responded. "Two close holes will create a vortex and water will pass through both of them faster than the sum of the speeds of each hole alone."

de Leon nodded in agreement. "Whatever the reason, that was the effect. The village elders were unhappy with that, because they had already established a full moon ritual in which percussion calabazas would be hit every minute as the village women chanted.

"They told the artisan that he must figure out how to make a 1-minute calabaza by the next full moon. He told them that he could not, but that with five calabazas, four assistants, a ready supply of water, and about an hour, he would be able to tell the drummers to hit their drums every minute exactly."

"How long does it take to fill an empty calabaza?" Liane asked.

"Natasha told me you were clever," de Leon said to the 11 year old. "Less than a second. One can just dip it into a big barrel. You can assume it takes no time."

"Well, then it shouldn't be so hard," Liane said. "For example, it's easy enough to measure 6 minutes. You start two calabazas at the same time, opening one hole of one and two holes of the other. Six minutes is the time between the draining of the two-hole calabaza and the draining of the one-hole calabaza."

"I think you are on the right track," de Leon said. "But it is important that you figure this out today. You see the full moon is about to come and the village artisan has taken terribly ill. They are counting on me to help them figure out what he did. They refuse to let me use my watch, as they revere the gourds as holy. Now for my personal appeal: As an ethnobiologist, I have made it my goal to win their confidence. They trust me so far and they have taught me about many wonderful herbs. They even have a remarkable medicine to enhance musical ability, seemingly without side effects, but I must stay with them longer before they will teach me about it. I would really like to help them with their problem first."

"Let me try a small example," Liane said. "Suppose that a one-hole calabaza drained in 3 minutes and a two-hole calabaza drained in 2 minutes [reversing the vortex effect for the sake of the example]. Then it would be possible to measure minutes by allowing calabaza A to drain in 3 minutes, B to drain in 2 minutes, and then repeatedly filling each calabaza and draining it with two holes. So, A would mark times 3, 5, 7, 11, 13, and so on, and B would mark times 2, 4, 6, 8, 10, 12, and so on. Thus, it would be possible to mark minutes beginning 2 minutes after the proecedure starts."

"Yes, yes!" de Leon said excitedly. "The artisan said something about marking time, but I didn't get it. Your idea should work but please do it for the five gourds the artisan carved."

Liane and Ecco talked for a while. Liane explained an idea. Ecco smiled.

"Mr. de Leon," Ecco said with a smile, "Liane has solved your puzzle. If five people follow this simple procedure, they will be able to mark single minutes starting 10 minutes after beginning the procedure."

1. Cybernovice, but hard: Can you invent a procedure that involves filling calabazas and watching them drain such that it is possible to mark single minutes starting 10 minutes into the procedure? Assume that it is easy to tell when a calabaza drains. Can you do it better (either with fewer than 5 calabazas or fewer minutes to start)?

Panta de Leon wrote down the solution rapidly and then left. To our great surprise he came back only 3 hours later.

"Something terrible has happened," he said. "A neighboring tribe stole three of the calabazas, thinking they were some kind of evil magic, so we have only two left. We feel lucky, though. The remaining two are so closely crafted that the chief calls them 'twins.' It seems that they hold exactly the same quantity of water."

"As a consequence," Liane said, "they must pour at exactly the same rate, too. I wonder if we can use that."

Liane and Ecco began to talk the problem over. As Liane explained her idea to him, it was not long before Ecco was nodding his head in agreement. Finally, Ecco turned around to face de Leon with a smile on his face.

"Mr. de Leon," he said, "Liane has found a procedure to mark minutes for the ceremony with just the twins."

2. Cybernovice, but hard: Can you figure out a way to mark minutes for the entire ceremony using only twin calabazas? What's the minimum amount of time that it takes to prepare the gourds for the ceremony?

3. Cybernovice, but hard: Suppose there had been no robbery and you have five gourds that are all "twins." Could you start the ceremony in 9 minutes instead of 10? Hint: consider making several gourds flow into one.

Solutions

1. Given an 11-minute gourd and two runs through a 5-minute gourd, you can construct a 1-minute gourd. In fact, if you start with four 11-minute gourds and one 5-minute gourd, you can construct four 1-minute gourds at the end of 10 minutes. At that point, you plug the holes of three of the 11-minute gourds and drain them one

at a time. Once a gourd is drained it is refilled but with both holes open, so each gourd marks 5 minutes but at 1-minute offsets.

	ACTIVITY	TIME LEFT (MINS)
Time 0:	A full, draining through two holes.	5
	B full, draining though one hole.	11
	C full, draining though one hole.	11
	D full, draining though one hole.	11
	E full, draining though one hole.	11
Time 5:	A empty; refilled; draining through two holes.	5
	B draining through one hole.	6
	C draining through one hole.	6
	D draining through one hole.	6
	E draining through one hole.	6
Time 10:	(ceremony begins)	
	A empty; refilled; draining through two holes.	5
	B draining through one hole.	1
	C plugged just now.	1
	D plugged just now.	1
	E plugged just now.	1
Time 11:	A draining through two holes.	4
	B empty; refilled; draining through two holes.	5
	C just unplugged; draining through one hole.	1
	D still plugged.	1
	E still plugged.	1

	ACTIVITY	TIME LEFT (MINS)
Time 12:	A draining through two holes.	3
	B draining through two holes.	4
	C empty; refilled; draining through two holes.	5
	D just unplugged; draining through one hole.	1
	E still plugged.	1
Time 13:	A draining through two holes.	2
	B draining through two holes.	3
	C draining through two holes.	4
	D empty; refilled; draining through two holes.	5
	E just unplugged; draining through one hole.	1
Time 14:	A draining through two holes.	1
	B draining through two holes.	2
	C draining through two holes.	3
	D draining through two holes.	4
	E empty; refilled; draining through two holes.	5
Time 15:	A empty; refilled; draining through two holes.	5
	B draining through two holes.	1
	C draining through two holes.	2
	D draining through two holes.	3
	E draining through two holes.	4

And so on. At this point, there is one gourd emptying every minute. The procedure is to keep refilling the newly dry gourd each minute and to drain it through two holes for as long as necessary—sounding the gong as the gourd runs dry each minute, of course. Ralph Fellows proposed this solution.

2. Use one twin as a 5-minute calabaza (two holes) and one as an 11-minute calabaza (one hole). Fill them both with water and drain both of them starting at the same time. The 5-minute calabaza will be the first to drain completely. It is refilled immediately while the 11-minute calabaza is still draining. Remember that we assume that refilling the calabazas takes no time so there is no time lost refilling the 5-minute gourd. When the 5-minute gourd drains the second time, immediately plug the one hole in the 11-minute calabaza. There is now 1 minute of water in the 11-minute gourd. Now you are ready for the ceremony. To keep on marking minutes throughout the ceremony, simply plug the holes in the 5-minute calabaza and let the 1 minute of water contained in the 11-minute gourd drain *into* the 5-minute calabaza. Then, unplug the 5-minute calabaza and allow it to drain into the 11-minute calabaza. Therefore, the same minute of water is used over and over again. You rotate in this way, switching the water from one gourd to the other, marking 1 minute with every switch, plugging and unplugging the appropriate gourds as you go. Therefore, the preparation and execution of the ceremony require only two calabazas, 10 minutes, and a very quick and dexterous person to perform the switch. Jimmy Hu and Scott J. Taylor proposed this solution.

3. The following approach from Magne Oestlyngen requires that several gourds can flow into one. His solution works assuming that the flow is constant from each gourd over time and across gourds. Let calabaza 1 (c1) be a 5-minute gourd and c2, c3, and c4 be 11-minute gourds. In his procedure c1 is allowed to empty and then is immediately refilled.

c5 is left empty at the start. Pipe the flows from c2, c3, and c4 into c5 (c5 has both holes open).

After 5 minutes (which end when c1 is first empty), $3 \times 5/11 = 15/11$ have flown into c5, and $5/5 = 11/11$ have flown out. That leaves 4/11 still in c5. Let c2, c3, and c4 then flow normally onto the ground. Plug one of the holes in c5.

So, c5 drains in 4 more minutes (at minute 9), whereas c1 drains the second time at 10. When c5 drains out, c1 drains its last minute into c5, marking minute 10 when it has finished; then, c5 drains into c1, and so on.

Inspiration and Offshoots

As a child, I always enjoyed puzzles that entailed pouring water from one bucket to another to arrive at some desired amount. This puzzle is a translation of that problem to time. Extensions of this problem are possible having more variants of flow rates or timings, but the elements of the solutions proposed by the readers showed considerable imagination. I particularly appreciated the idea of exploiting the possibility that two calabazas could hold the same volume.

21

Subway Scents

Police commissioner Bratt returned. His entourage this time consisted of only two individuals, whom he introduced as "my colleagues Kris and Joe." What a pair! The young woman's snug black dress, silky black hair, black beret, and dark sunglasses set off her bright red lipstick. She tilted her head slightly as a greeting and then sat down lightly on the chair, with no intention of relaxing her posture. The man wore jeans held together by a toolbelt, on which were hanging a flashlight, a measuring tape, and a set of screwdrivers. His fingernails showed stains of machine oil. He sat heavily into an armchair, visibly exhausted. Then he pulled a pad out of his back pocket.

"Joe here is a subway engineer," Bratt explained. "In some ways, he is *the* subway engineer for our great city in that he makes sure the trains run on time."

"To the extent they do at all," Joe said with a disarming smile.

"Kris is, well, let's say a strategist," Bratt went on.

Liane spoke up, her voice laden with sarcasm: "Versace, aren't they?" she said pointing to Kris's sunglasses.

This adventure occurred in November 1998.

"That's right, Liane," said Kris turning her head toward the 10 year old. "They contain directional listening devices. I can pick up conversations at 50 feet in a crowded train or in the bar of a hotel."

Liane staggered backwards a few steps, her eyebrows raised.

Kris followed Liane with her shaded gaze and continued, "You and your uncle helped the commissioner greatly in his drug-fighting efforts. You even helped convince him that I was an honest informant . . . though he prefers nonmathematical ways to determine honesty."

I could just imagine.

"Now, we need your help to solve a problem of, well, distribution. Will you help us?" Kris asked, still looking at Liane.

"Sure," said the young girl, far from recovering her composure. "Anything you want."

"Absolutely anything," Ecco said with a smile.

"Very well," said Bratt. "Joe, please show the map."

Joe spread out a map for us (see Figure 1). "The midtown stops of the subway resemble a grid and spread from upper Brooklyn to lower

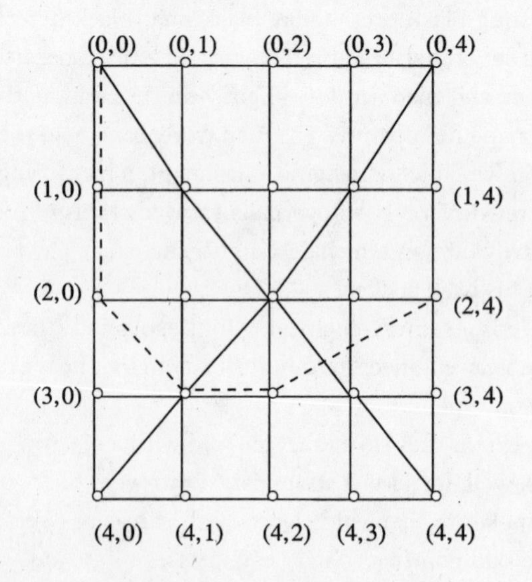

Figure 1. Midtown subway stops.

Queens and midtown Manhattan," he explained. "I'm abstracting slightly, but topologically it's a 5 by 5 grid with a few extra lines. Each line represents a pair of tracks. [For concreteness, label the stations, including the interior ones, from (0, 0) in the top left corner to (4, 4) in the bottom right, as you see in the figure.]

"At each end station—the 16 around the periphery—a train begins at 6 AM with two exceptions: The horizontally moving trains starting at (2, 0) and (2, 4) begin at 6:20. At 6 AM, there is a lot of activity in the corners. Each of three corners has three trains leaving—one horizontally, one vertically, and one diagonally. The northwest corner has four trains leaving at 6 AM. Train stations are 10 minutes apart. It takes 10 minutes for a person to switch trains at a station. Trains turn around at the terminus stations in negligible time.

"You notice the diagonal lines and the line I've drawn with dashes. Trains go slightly faster on those lines to maintain the 10-minute station-to-station pace. The dashed line between (0, 0) and (2, 4) hits stops at (0, 0), (1, 0), (2, 0), (3, 1), (3, 2), and (2, 4). It's our best line, by the way. It runs at 30,000 volts higher than the other lines to conserve power and the brake voltage ..."

"Thank you, Joe," Bratt interrupted. "Now, Dr. Ecco, let me explain the scenario. Suppose a nasty person could start at any station he desired at some time and wanted to deposit, well, smelly postcards in as many stations as possible in as short a time as possible..."

Kris took over, "How long would it take that person to drop postcards off at eight stations assuming he started at 6 AM?"

"Including the starting station?" Liane asked, forgetting her nervousness now that a puzzle was posed.

"Correct, Liane," Kris replied.

"Well, then he could start at 6 AM at station (0, 0) and drop a postcard there, then go to (1, 0), (2, 0), (3, 1), (3, 2), and (2, 4), dropping cards as he went. That takes 50 minutes. Then he could switch to the train that started at (0, 4) and went down, but is now on its way back up. After 50 minutes, it's at (3, 4). It takes him 10 minutes to switch, so he'd be on it at 60 minutes and would be able to go to (1, 4) and (0, 4) in the next 20 minutes. That makes 7 stations in 80 minutes."

"Nice job, girl," said Joe, smiling with delight. "We've come to the right place."

Kris continued, "Right, except we knew that answer. Here's one we don't know. How long would it take a nasty person to drop postcards off at 13 different stations [12 besides the starting station]? You can start at any station and at any time."

1. Cybernovice: Can you find a solution that works in 140 minutes?

When she heard Liane's clever answer, Kris smiled, but it was fleeting. Bratt went on to the next question. "Suppose the person targeted 19 stations. Can he do that in under 300 minutes?"

"Quite a bit under," said Ecco after a few minutes thought. "Here is how."

2. Cybernovice: Can you have the nasty person hit 19 stations in 240 minutes or less?

Lifting his head up from the notes he was taking, Joe asked, "What is the maximum number of stations that two nasty people can distribute postcards to in 80 minutes?"

"I would say at least 15," said Ecco. "Liane, what do you think?"

3. Cybernovice: Liane thinks 16 may be more like it. What do you say and how would you do it?

After staring at the sheet of paper in silence for a while, I volunteered a new observation, "Hmm. And five nasty people can make all your subways stink in less than an hour."

4. Cybernovice: Show how (this one is easy).

After listening to my explanation, Joe went over his notes. Then he turned to Bratt. "Commissioner, we have a big problem. Teams should be ready at every stop."

Solutions

1. Alan E. Dragoo proposed a method for dropping 13 stinky postcards off in 140 minutes after getting some extra sleep (and, ominously, acting closer to rush hour):

Board the special train at station (2, 4) at 7:40 AM, arriving at (0, 0) at 8:30 AM. The perp then boards the eastbound train at 8:40 and disembarks at (0, 3) at 9:10. Finally, he/she boards the southbound train at 9:20 and arrives at (4, 3) at 10:00.

2. Alan also sent in one of many good 19 station/240 minute solutions:

Starting at (0, 0) at 6:00, he/she takes the special train to (2, 4) at 6:50 AM; takes the 7:00 westbound train to (2, 1) at 7:30; switches to the northbound train at 7:40, arriving at (0, 1) at 8:00; 8:10 east to (0, 3) at 8:30; 8:40 south to (4, 3) at 9:20; and 9:30 west to (4, 0) at 10:00.

3. If two nasty people work for 80 minutes, they can drop off 16 cards. The first nasty person takes the route that Liane suggested: (0, 0), (1, 0), (2, 0), (3, 1), (3, 2), and (2, 4), then switches to (1, 4) and (0, 4). The second nasty person takes the route (0, 3), (1, 3), (2, 3), (3, 3), and (4, 3), then switches in 10 minutes to (4, 2), (4, 1), and (4, 0).

4. Professor Scarlet's observation is easy to show: Five people can cover every station in 50 minutes, for example, if they start from

(0, 0), (0, 1), (0, 2), (0, 3), and (0, 4), respectively, and go down in parallel.

Inspiration and Offshoots

Here the mathematics require simulating in one's mind or on a computer the future trajectories of the subways and trying to find the times when the most stations will be visited in the shortest time. This is a special case of forecasting that is also used in the Escape puzzle. Luckily, the New York subways are not very good at keeping to their schedules.

22

Stars, Money, and Time

The sunglasses, beach shirt unbuttoned nearly to his well-fed bellybutton, and golf tan should have given him away. Joe Rouge was a movie producer. He had bought into his first movie after making a killing in fast food real estate.

"You gotta see it to believe it, Doc," he said to Ecco, waving his unlit cigar around for emphasis. "These expensive actors just sit around most of the time at these movie sites. They're not rehearsin', they're sittin' watchin' a scene get set up that they're not even in. Now, I'm no mathematician, but I got to thinkin': Can't we move them around so they don't waste so much time and so I don't gotta pay them? I asked my nephew Gary here to help me out, but he seems to think that only you can solve the problem right. This movie is costin' me a bundle. If you can reduce the expenses even by a million, I'll pay you handsomely."

Gary—pale, stick thin, his narrow shoulders hunched forward slightly—stood up and spoke in a soft voice. "Here is how it works. Actors in our studio are paid by the day. The three stars Patt, Casta, and Scolaro are paid exceptionally well, several hundred thousand dollars a day in fact. Each of the 19 scenes requires its own cast, and setting up a scene and shooting all its takes requires a long time. So, we figure five scenes a day. We are currently filming in one location—a cushion-rich mansion with Pacific Ocean backdrops, if you get my drift—so the scenes can be ordered any way you want."

"The crew will be on scene all the time and at the same expense?" Liane asked.

Gary seemed startled at the sound of the 11-year-old girl, but answered, "Yes, the crew cost and equipment rental is constant but is very high, about $10,000 per hour in fact."

Gary paused, as if unsure of what to do next. Rouge tapped his cigar impatiently on the edge of a chair. Liane spoke up, "Gary, could you tell us the current order and composition of the scenes and then the daily rates of the actors?"

"Yes, of course," Gary replied, reaching into his suitcase.

"Here's a list of the actors in each scene, listed in the order of the script. Each line represents one scene.

Hacket;
Patt, Hacket, Brown, Murphy;
McDougal, Scolaro, Mercer, Brown;
Casta, Mercer;
Mercer, Anderson, Patt, McDougal, Spring;
Thompson, McDougal, Anderson, Scolaro, Spring;
Casta, Patt;
Mercer, Murphy;
Casta, McDougal, Mercer, Scolaro, Thompson;
Casta, McDougal, Scolaro, Patt;
Patt;
Hacket, Thompson, McDougal, Murphy, Brown;
Hacket, Murphy, Casta, Patt;
Anderson, Scolaro;
Thompson, Murphy, McDougal, Patt;
Scolaro, McDougal, Casta, Mercer;
Scolaro, Patt, Brown;
Scolaro, McDougal, Hacket, Thompson;
Casta.

"Here are the daily rates of each actor:

Patt: $264,810;
Casta: $250,430;
Scolaro: $303,100;
Murphy: $40,850;
Brown: $75,620;
Hacket: $93,810;
Anderson: $87,700;
McDougal: $57,880;
Mercer: $74,230;
Spring: $33,030;
Thompson: $95,930.

"This gives the following total expense, excluding the crew: $4,975,360. With the crew, the expense is over $5 million." Gary paused, then looked over his shoulder at Mr. Rouge who was no longer paying attention. "This is a very B movie, please understand," Gary whispered. "5 million is really too much."

Ecco studied the figures while he was munching on an oatmeal cookie. "Mr. Rouge, we'll have to study this for a day," he said. "Could you come back tomorrow?"

"Really?" Rouge asked, thinking he detected a note of reluctance on the part of Ecco. "Doc, whatever you save me from the $5 million, I'll pay you 10% of that amount." He looked at Ecco, searching for some sign of greed.

"That's fine," said Ecco, with a friendly grin. "Until tomorrow then."

They came in the next day and Rouge barely greeted us. "What did you find out, Doc?"

"Tell him, Liane," Ecco said.

"Well, Mr. Rouge," she said, presenting a piece of paper to the producer. "We can save you over $1,600,000 with the following organization."

Advanced cybernovice: Can you do as well or better?

"That's great, Doc," said Mr. Rouge. "How much did I say I'd pay you?"

Solution

Try the following ordering of the scenes.

	SCENE	ACTORS
Day 1	1	Hacket
	6	Thompson, McDougal, Anderson, Scolaro, Spring
	14	Anderson, Scolaro
	18	Scolaro, McDougal, Hacket, Thompson
Day 2	3	McDougal, Scolaro, Mercer, Brown
	9	Casta, McDougal, Mercer, Scolaro, Thompson
	10	Casta, McDougal, Scolaro, Patt
	16	Scolaro, McDougal, Casta, Mercer
	17	Scolaro, Patt, Brown
Day 3	2	Patt, Hacket, Brown, Murphy
	5	Mercer, Anderson, Patt, McDougal, Spring
	11	Patt
	12	Hacket, Thompson, McDougal, Murphy, Brown
	15	Thompson, Murphy, McDougal, Patt
Day 4	4	Casta, Mercer
	7	Casta, Patt
	8	Mercer, Murphy
	13	Hacket, Murphy, Casta, Patt
	19	Casta

This gives a final cost of $3,341,440.

This solution was first found by Jon Beal, Thomas Cloutier, Tomas G. Rokicki, Jason Strickland, and Onno Waalewijn.

This is a classic search problem whose guaranteed best answer can be found only by searching all possible permutations. However, there are many good problem-solving methods (heuristics) such as to take the expensive actors and try to cast them on the same day. That was actually Liane's strategy.

Inspiration and Offshoots

I walk my kids to school in Greenwich Village. Roughly once a week, we pass a movie set. To my untrained eye, nearly everyone on the set is doing, well, nothing. They are waiting, while trying to look important. Someone is paying them. From this, one may wonder which stars are paid to wait and which ones are scheduled right away. This problem is in the class of computer science known as NP-complete problems, meaning roughly that the cost of a solution is easy to compute, but a solution of a given low cost may be time-consuming (exponential in the size of the problem) to find. But for life to be interesting, it ought to be challenging.

23

Trains for the Sultan

It was unusual for Ecco to answer questions posed by 4 year olds, but that was what he was being asked to do. The mouthpiece, as it happened, was the Sultan of Brunei. This gentleman, possibly the richest in the world, draws his wealth from a lake of oil that sits under his country's jungle. His palaces, some larger than Versailles, dot the land. His subjects pay no taxes, get free medical care, and receive free television sets. Wealth has other privileges. For example, the Sultan bought the Dorcester hotel in London, because he arrived at the hotel one night unannounced and was told there were no rooms available. Purchasing the hotel has vastly improved his chances of getting a room on demand it seems.

During a recent visit to the United States, he asked Ecco to come to his suite in the Plaza hotel to solve a design problem. Ecco, in turn, asked Liane and me to come along.

In spite of his wealth, the Sultan conveyed an air of disarming informality. He did not remark on our untailored clothes or the fact that we were three instead of one, seeming to be well aware of our working style. Wanting at once to dispel any feeling of unequal status, he stood up and shook our hands when we entered. Tea was served to all. We sat around an elegant antique desk in equally comfortable chairs.

His secretary of protocol had requested only one favor from Ecco when he called to invite him: "Please let the Sultan ask all the questions. Commoners are not to ask questions of royalty, even honored commoners like yourself."

With that admonition in mind, we waited for the Sultan to begin the conversation. The Sultan approached the problem indirectly, "As this young lady will certainly appreciate," said the Sultan pointing to 10-year-old Liane, "children can come to a nearly fanatical early fascination with objects. So it is with my 4-year-old son Hasan. He has a fine collection of electric trains which he plays with for hours on end, a concentration befitting a much older child. He memorizes train maps and has his governess quiz him on them. He looks at my fleet of Rolls Royces with disdain, preferring the da-da-da-dun of rolling stock.

"Two weeks ago, he put the question to me this way: 'Father, why are there so many more cars than trains?' I thought about this awhile and then responded, 'Because cars go wherever you ask them to, whenever you ask them to. Trains have fixed routes on fixed schedules.' Hasan fell silent. A day later, though, he came back to me: 'Father, in our beloved Brunei, if we built tracks where we now have roads, then trains would be able to go to most places where cars go. That would eliminate the problem of fixed routes. As for the problem of fixed schedules, we could make the trains come at our demand.' "

"He's very smart for a 4 year old," Liane said appreciatively.

The Sultan nodded and smiled. He then raised his index finger and more tea was served. "Little Hasan is also single-minded," the Sultan continued. "He has mapped out the routes among the seven principal population centers of Brunei. The names are difficult for Westerners to pronounce, so I have designated them with letters for you. The letter B represents our downtown and the letter E represents a principal recreational area."

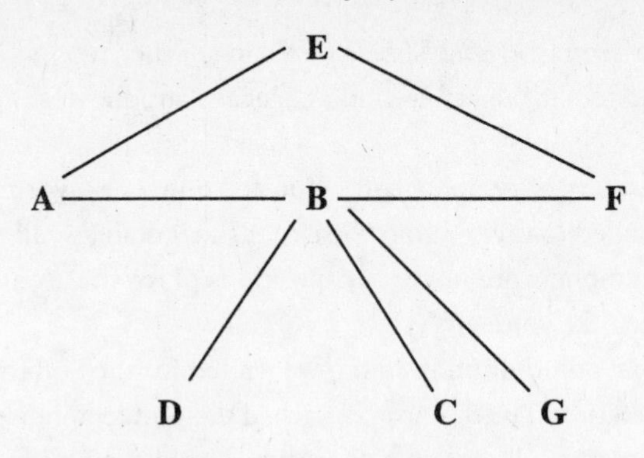

Figure 1. Each of the letters stands for a train station. The line segments represent single tracks that trains can traverse in either direction, though only one direction at a time. Each line segment takes 15 minutes to traverse. Initially, there is one train in each station, though there is room for two. The time for passengers to get on or off a train is negligible compared to the travel time.

The three of us studied the map (see Figure 1).

"My engineers suggest a light rail system where each train has three cars and can carry 50 people in the comfort they are accustomed to. They have designed the system so it takes 15 minutes to traverse a leg from one station to its neighbor. Each train station will have room for two trains, but there will be only one track between stations. The engineers say that there is no justification for more tracks. Several trains can go on the same track, provided they are going in the same direction.

"Little Hasan, on the other hand, wants every passenger to get to his or her destination without changing trains and by one of the shortest paths possible. Further, he wants to be sure that even at times of peak demand, passengers can get to their destinations quickly.

"Here is the demand at rush hour:

150 want to go from A to F;
100 want to go from B to E;
50 from C to G;

150 from C to D;
150 from D to C;
100 from E to B;
50 from E to D;
50 from F to C;
50 from F to A;
50 from G to A.

"Can you, Dr. Ecco, ensure that everyone arrives at his or her destination as fast as possible, given that," the Sultan put on his reading glasses and read a summary of all the conditions, "there is initially a train at each destination, each leg takes 15 minutes to traverse, there is only one track between stations, trains can take only 50 people at a time, no passenger needs to switch trains, and no passenger spends more time once on a train than the minimum time necessary?"

1. Cybernovice: Please have a try. You should be able to do it all in less than 100 minutes.

Liane and Ecco began working immediately. After several minutes, they presented a schedule and train routing to the Sultan in which the latest passenger arrived at his or her destination after less than 95 minutes.

The Sultan nodded. "That is very good, but I think Hasan will not be pleased," he said. "It's just too much time. Is there any engineering change that would lead to a faster solution?"

"If you convert five of your trains to carry 150 people instead of 50, allow station B to hold four trains instead of two, buy one more 50-person train, allow me to change the starting position of one of the existing trains, and allow intermediate stops, then we can get everyone to his or her destination in 45 minutes," Liane volunteered.

"Please tell me how," the Sultan said.

2. Cybernovice: Can you match Liane's achievement or find a more efficient way? More efficient means either reducing the station size of B or the train size of at least some of the trains.

After she presented her solution, the Sultan beamed. He snapped his fingers and a servant arrived with a glistening diamond necklace.

"For a future occasion," he said to Liane as he presented it to her.

Solutions

1. When trains can carry only 50 people, James Waldby found a solution requiring only 90 + 3 epsilon minutes for all of rush hour, where epsilon is the time to wait. An asterisk represents that epsilon time. For convenience, let's make epsilon be 1 minute. Typical trains do 2-minute stops. In his notation, consecutive letters without spacing represent a trip without stops at the intermediate stations (e.g., AEF represents an express trip from A to F without stopping at E).

START TIME	ROUTE TAKEN
0	AEF * FBC CBG
0	BAE EFB BAE
0	CBD DBC ** CBD
0	DBC CBD ** DBC
0	EFB BA AEF FB
1	FBA AEF FB BA
2	GBA AE * EFBD

2. Chris Rosenbury proposed the following when five trains can carry 150 passengers and are allowed to make several stops. The goal is to reduce the size of station B to two platforms. His basic idea was to stagger train departures.

START TIME	ROUTE TAKEN	NUMBER OF PEOPLE ONBOARD
0	EFBD	150, taking care of both E to B and E to D.
0	AE, wait 15 minutes, then EF	150
0	FBAE	150
0	GBA	50
5	CBG	50
10	CBD	150
10	DBC	150
15	FBC	50

Inspiration and Offshoots

Moving people from here to there is a constant obsession in big cities. Most industrialized countries aside from the United States achieve this mostly by rail. Many of those countries lay new tracks, expand rail stations, and undertake other infrastructure efforts to discourage people from driving and thereby sitting in polluting traffic jams or consuming the downtown with parking lots. The question this puzzle poses is how to best allocate resources and under which schedule. The question must be entirely reconsidered if the trains could change their destinations depending on the instantaneous demands of the passengers, for example, an upcoming sports event or party sponsored by the Sultan. To be concrete, suppose that the Sultan asked Liane to consider what would happen if the network included a special stadium site S having links to A, B, and F. Normally, no traffic goes to S, but when a special event occurs, three extra trains are introduced. In that case, how little could Liane disrupt her best schedule while guaranteeing that 420 people could go to the stadium in the space of an hour, 60 from each other site?

24

Fast Shipping through Panamax

With her wispy hair and dreamy eyes, she looked the poet she was in her spare time. In her day job, however, she was a shipping magnate and Katy McLean had oceanic ambitions. "In 1946, my granddaddy Malcolm McLean sent a metal container from New York City to a southern U.S. port," she began. "Now he always thought big, but how could he know that this little innovation would turn the shipping world upside down and bring fabulous profits to his little company that became Sea-Land?

"Here I am trying to extend the adventure with a fleet of fast, specially designed panamax ships. Postpanamax ships are too large to pass through some of the smaller locks in the Panama Canal. They carry a lot but the trip around the tip of South America is quite a detour. Current rules limit ships to carrying around 13 piles of containers, each around nine deep. We want a ship long enough to hold only four piles, but 12 deep, so it will fit comfortably in the canal, even allowing two-way traffic, and still fall within the canal's 200-foot height limit. This will speed our journey. Our technology allows us to hold our 10-foot-high containers even when sailing over rough ocean. Building the ship is no problem, but what is hard is loading and unloading her efficiently. To keep our ships busy, we will send our ships to eight of the busiest ports in the world: South Louisiana, New York, Le Havre, Rotterdam, Singapore, Shanghai, Hong Kong, Los Angeles, and then back to South Louisiana. Docking fees are high, though, so we want to reduce them to a minimum by minimizing the number of containers that need to be moved on and off a ship each time.

"To give you some idea what is going on, say we have four ports visited in the order (A, B, C, D, A. . .) and two piles. One container is added at each port and goes to the farthest possible port. So, at A a container is added going to D. At B, a container is added going to A, and so on. Containers are removed at their destination. Let's see how this might go:

At Port A: (D), ( );
At Port B: (D), (A);
At Port C: (D), (B,A), where (B,A) means B on top of A;
At Port D: Remove D and put on C to get (C), (B,A);
At Port A: Move B (overhead) to get (B,C), (A);
then, remove A and put on D to get (B,C), (D);
At Port B: Remove B and put on A to get (C), (A,D);
At Port C: Remove C and put on B to get (B), (A,D);
At Port D: Move A (overhead) to get (A,B), (D); then,
remove D and put on C to get (A,B), (C);
At Port A: Remove A to get (B), (C); then, move B
to get ( ), (B,C); finally, add in D to get (D), (B,C).

"So, there is an overhead of two moves per cycle. Each move costs us time, and time means money, Dr. Ecco."

"How much time does a move cost?" Liane asked.

"That depends on the port," McLean responded, unfazed that a 13 year old was asking her questions, "maybe only a few minutes. It doesn't matter. Dock time is expensive, so we want to minimize it."

"Can you put a crane on the ship and move containers while the ship is in the ocean?" Liane continued.

"Good question," McLean responded. "We thought of that, but decided against adding the cranes, because we didn't believe and still don't believe we can guarantee stability with piles 12 high."

"OK, so each move costs a certain amount of time," Ecco concluded. "Can the dock area near the ship be used, say, to rearrange an entire pile?"

"On our ships, yes," McLean answered. "So a 12-container pile can be completely reorganized in 24 moves. In fact, we suspect this might be useful, at least for partial rearrangements."

"OK, Liane, let's try a new problem," Ecco said. "The ship is going to eight ports [numbered 0, . . . , 7] and has two piles of up to four each. Every port deposits one container that goes to the farthest possible port. What is the smallest overhead (extra moves) per cycle that you can manage?"

> 1. *Cybernovice: Liane was able to manage with an overhead of 22. Can you do better?*

McLean liked the solution Liane came up with. "Now for the real problem," she said. "You have the eight busy ports I mentioned before and four piles each of height 12. Each port sends six containers to one or several other ports. What is the smallest overhead per cycle that you can manage for sure? Is the most expensive case always when the containers are sent to the most faraway ports [in the cycle, not geographically]?"

> 2. *Cybernovice: Would you like to try your hand at these two questions? Dr. Ecco and Liane were working on them when I left.*

Solutions

1. Steve Hammer proposed a solution having an average of $13\frac{1}{3}$ overhead moves per cycle and whose initial configuration was repeated every three cycles. The overhead varies per cycle from 10 to 15 and again 15. Start with a "full" (seven containers), well-ordered ship. OH means overhead and the parentheses denote the order of containers in the two piles.

At Port 0: (1,2,3,4), (5,6,7);
At Port 1: Remove 1 and put on 0 to get (2,3,4), (0,5,6,7);
At Port 2: Remove 2 and put on 1 to get (1,3,4), (0,5,6,7);
At Port 3: Remove 1 (OH), remove 0 (OH), remove 3, and move 4 (OH) to get ( ), (4,5,6,7); then, put on 2, put on 1 (OH), and put on 0 (OH) to get (0,1,2), (4,5,6,7);
At Port 4: Remove 4 and put on 3 to get (3,0,1,2), (5,6,7);
At Port 5: Remove 5 and put on 4 to get (3,0,1,2), (4,6,7);
At Port 6: Remove 4 (OH), remove 3 (OH), remove 6, and move 7 (OH) to get (7,0,1,2), ( ); then, put on 5, put on 4 (OH), and put on 3 (OH) to get (7,0,1,2), (3,4,5);
At Port 7: Remove 7 and put on 6 to get (0,1,2), (6,3,4,5);
At Port 0: Remove 0 and put on 7 to get (7,1,2), (6,3,4,5);
At Port 1: Remove 7 (OH), remove 6 (OH), remove 1, and move 2 (OH) to get ( ), (2,3,4,5); then, put on 0, put on 7 (OH), and put on 6 (OH) to get (6,7,0), (2,3,4,5);
At Port 2: Remove 2 and put on 1 to get (1,6,7,0), (3,4,5);
At Port 3: Remove 3 and put on 2 to get (1,6,7,0), (2,4,5);
At Port 4: Remove 2, remove 1, remove 4, and move 5 (OH) to get (5,6,7,0), ( ); then, put on 3, put on 2 (OH), and put on 1 (OH) to get (5,6,7,0), (1,2,3);
At Port 5: Remove 5 and put on 4 to get (6,7,0), (4,1,2,3);
At Port 6: Remove 6 and put on 5 to get (5,7,0), (4,1,2,3);
At Port 7: Remove 5 (OH), remove 4 (OH), remove 7, and move 0 (OH) to get ( ), (0,1,2,3); then, put on 6, put on 5 (OH), and put on 4 (OH) to get (4,5,6), (0,1,2,3);
At Port 0: Remove 0 and put on 7 to get (7,4,5,6), (1,2,3);
At Port 1: Remove 1 and put on 0 to get (7,4,5,6), (0,2,3);
At Port 2: Remove 0 (OH), remove 7 (OH), remove 2, and move 3 (OH) to get (3,4,5,6), ( ); then, put on 1, put on 0 (OH), and put on 7 (OH) to get (3,4,5,6), (7,0,1);
At Port 3: Remove 3 and put on 2 to get (4,5,6), (2,7,0,1);
At Port 4: Remove 4 and put on 3 to get (3,5,6), (2,7,0,1);

At Port 5: Remove 3 (OH), remove 2 (OH), remove 5, and move 6 (OH) to get ( ), (6,7,0,1); then, put on 4, put on 3 (OH), and put on 2 (OH) to get (2,3,4), (6,7,0,1);
At Port 6: Remove 6 and put on 5 to get (5,2,3,4), (7,0,1);
At Port 7: Remove 7 and put on 6 to get (5,2,3,4), (6,0,1);
At Port 0: Remove 6 (OH), remove 5 (OH), remove 0, and move 1 (OH) to get (1,2,3,4), ( ); then, put on 7, put on 6 (OH), and put on 5 (OH) to get (1,2,3,4), (5,6,7).

2. James Waldby proposed a 24-move overhead solution to the second problem: You have the eight busy ports I mentioned before and four piles, each of height 12. Each port sends six containers to the farthest port in the cycle. What is the smallest overhead per cycle that you can manage?

Here is Waldby's solution, where each six-container group is denoted by a single letter (the letter of its port of destination) and stacks are listed from bottom to top (so EA means A over E):

PORT	ARRIVAL STACKS	COST TO LEAVE	ACTIONS TO LEAVE
A	FC EA DB G	0	off A, on H
B	FC EH DB G	0	off B, on A
C	FC EH GA D	0	off C, on B
D	FB EH GA D	6	off D, on C, move H
E	CH FB GA E	6	off E, on D, move B
F	DB CH GA F	0	off F, on E
G	DB CH GA E	6	move A, off G, on F
H	DB CH EA F	6	off H, move C, on G
	Total	24	

The question remains open whether sending containers to the farthest port yields the highest overhead.

Inspiration and Offshoots

The best way to transport cotton is by slightly rounded bale, not by container. The best way to send a house's belongings is not by container, either. Containers waste volume. Yet containers have taken over shipping because they have three advantages. They can move on ships, trains, and trucks without repacking their contents. Manufacturers build ships, cranes, train cars, and trucks to accommodate the same-shaped payload. Because containers are rectanguloids, they can be stacked. What may be suboptimal in one leg for one customer becomes optimal when considering many shipments by many customers—a kind of socialism at sea. The purpose of this puzzle is to explore how best to stack containers depending on their final destination. In general this is known as an "online" optimization problem, because the destinations of containers are discovered way after the ship is built. The puzzle as it stands has explored only some of the simple cases. Imagine a fleet of boats having different routes and containers having different deadlines. It gets very interesting very fast.

Part VI

Ciphers and Secrecy

25

Smuggling for a Greater Good

A few minutes after I arrived in Ecco's apartment, the doorbell rang. A slim, fit woman in her late twenties entered, introducing herself simply as Natasha. "My last name need not concern you," she explained as she looked us over, sizing us up with her penetrating blue eyes.

"I've come to ask for your help, Dr. Ecco," Natasha went on. She didn't sit down, though she did balance her briefcase on the back of a chair. "But first I want to give you a little background."

"We are ready," Ecco said with a smile.

"Heinrich Schliemann, the German grocer turned archaeologist, followed the *Iliad* to discover Troy. Later he thought he found Agamemnon's tomb in Mycenae," Natasha began. "He was wrong about Agamemnon, but he did find five tombs sculpted in gold, so they probably weren't shepherd graves. His excavation techniques were controversial, to say the least. He didn't shy away from dynamite if a hill was in the way. He also neglected to share the treasures of the Trojan jewel factory with the Ottoman Empire in spite of his promises to the Sultan. The Ottomans protested and demanded

10,000 gold francs. He gave them 50,000 instead. He was a man in a rush.

"Well, I'm a modern archaeologist, so you'd expect me to be a brush and silk kerchief kind of digger. Usually, I am, but not this time. You see, I'm excavating in the Sudan. The country is in a state of civil war and the authorities are deeply corrupt. So, I'm a woman in a rush. I'm going back tomorrow and I expect my team there to have found 12 large specimens. We want to smuggle them out of the country a few at a time and keep them until the Sudan has a responsible government."

"You're a thief," 10-year-old Liane said, ever tactful.

Ecco nodded, "Quite possibly, Liane. How do we know, Natasha, that you're telling the truth?"

Natasha looked unperturbed and reached into her briefcase. "You don't know I'm telling the truth, but here is a stack of my publications. I am known as a great preserver. The specimens will in fact go into a private collection, because the U.S. government doesn't want problems with the government of the Sudan over the issue. But I assure you we are doing this only to preserve the finds, not to exploit them."

"Why can't you just leave them in the ground?" Liane asked.

"The site is exposed now," Natasha said. "We are convinced the site will be looted if we delay."

"What can we do for you then?" Ecco asked without enthusiasm.

"Well," Natasha replied. "As I said, there are 12 possible finds. We want to identify each of them with a positive whole number. When we phone our home base we want to use the sum of the numbers corresponding to the finds to tell our colleagues which ones to prepare for. They may respond with other sums to tell us which ones they are ready for. We think it's better to send a sum rather than a sequence of numbers because it will attract less attention."

"Please give us an example," Ecco said.

"Suppose, for example, that there were only three possible finds and we labeled them 10, 11, and 12, respectively," Natasha replied. "Then, for example, 22 represents the objects labeled 10 and 12, whereas 33 represents all three objects. Altogether the following sums are possible, each of which represents a unique set of objects: One object could be 10, 11, 12; two objects could be 21, 22, 23; three objects could be 33 (10 + 11 + 12). On the other hand, if we had four possible finds, the encoding 10, 11, 12, 13 wouldn't work because 23, for example, could result from either 10 + 13 or 11 + 12. We want every possible sum to mean a different combination."

"That's easy enough if we can use big numbers," Liane said. "For example, 10, 100, 1,000, 10,000."

"Correct, but that's just the thing," Natasha replied. "We want the largest possible sum to be as small a number as possible. If someone hears us talking about numbers in the billions, they may think we are bank robbers."

"Big difference," Liane said in a loud whisper.

"But if every subset of 12 is possible, then the highest possible sum must be over 4,000," Ecco said.

1. Cybernovice: Why should this be?

"I see," Natasha said after she heard his argument. "That's OK in fact because these finds are so big, we might not be able to take out more than three at a time. Can you find a labeling that would guarantee us a low sum?"

Ecco found a way to give positive whole number labels to the 12 objects such that any subset of the objects of size 3 or less could be described by the sum of the labels and that sum would uniquely identify the subset.

2. Advanced cybernovice: Try to find a labeling giving the smallest maximum sum in that case. Dr. Ecco was able to find a labeling such that the sum of any three never exceeded around 1,050.

After hearing this answer, Natasha looked preoccupied. "Come to think of it," she said. "We may be able to get four specimens out at a time."

3. Cyberexpert: What would be the smallest maximum sum in that case? Liane was able to find a labeling that ensured that the sum of any four or fewer items was under 2,600.

Natasha asked a few questions, made some notes on the written solutions, and left.

Ecco turned to Liane and me, "I understood why Natasha wouldn't want fractions and I suppose she wouldn't permit negative numbers to be reported over the phone, but I wonder if we could find a way to use negative numbers to good effect. Maybe some labels could be positive in some cases [e.g., if the sum it participates in would be negative otherwise] and negative in others. I wonder whether we could get a better solution in that case."

4. Advanced cyberexpert: If negative labels are allowed and labels can be positive or negative depending on the context, can you achieve better results for the above two questions?

Ecco never told me whether this idea in fact helped.

Solutions

1. If every subset is possible, the total number of nonempty subsets of n elements is $2^n - 1$. Each of these must have a unique sum. If $n = 12$, then $2^n - 1 = 4{,}095$.

2. The best solution comes from reader Benjamin C. Chaffin who used the following encoding for sending any 3 of 12 objects: 2, 4, 8, 16, 32, 60, 73, 116, 207, 230, 341, and 452, yielding a maximum sum of 1,023.

3. He also came up with the best encoding when four objects could be shipped at once: 16, 18, 19, 20, 24, 32, 64, 128, 237, 420, 712, 1,187, with a maximum sum of 2,556.

4. On first glance, it might seem that negative numbers wouldn't help but they do. The basic algorithm is to use negative numbers and then to take the absolute value of any resulting sum. This guarantees a nonnegative answer, of course. The real cleverness is to ensure unique decodability. Dan Hirschberg showed that when at most three choices can be made, then negative numbers can reduce the maximum sum to 734: 2, 3, 4, 8, 16, 30, 56, 110, 173, 244, 317, −626. In the best solution, Ted Alper then showed that negative

numbers can reduce the maximum sum of 1,527 for up to four choices: 2, 3, 4, 8, 16, 32, 61, 116, 224, 416, 771, −1,468.

Since I know of no proof of minimality for any of these sequences, a clever reader may find a better solution than the ones above. For that reason, I've written a checking program in K that you can find on the Norton Cyberpuzzles Website, www.wwnorton.com/drecco. The program takes a sequence and the number of choices that can be made and sees whether the sequence results in the unique decodability of subset sums.

Finally, Jimmy Hu suggested encoding each set of numbers by a single number and simply sending the encoding. This violates the rule that the sum should be used, but I mention it because of its cleverness. So, given the encoding:

$$
\begin{aligned}
\{1\} &= 1,\\
\{2\} &= 2,\\
\{3\} &= 3,\\
&\vdots\\
\{12\} &= 12,\\
\{1,2\} &= 13,\\
\{1,3\} &= 14,\\
\{1,4\} &= 15,\\
&\vdots\\
\{11,12\} &= 78,\\
\{1,2,3\} &= 79,\\
\{1,2,4\} &= 80,\\
\{1,2,5\} &= 81,\\
&\vdots\\
\{10,11,12\} &= 298,\\
\{1,2,3,4\} &= 299,\\
\{1,2,3,5\} &= 300,\\
\{1,2,3,6\} &= 301,\\
&\vdots\\
\{9,10,11,12\} &= 793.
\end{aligned}
$$

If the desired set were 1, 2, 3, 6, then the label would be 301.

Inspiration and Offshoots

If you add $5 + 2$ or $4 + 3$, you get the same answer—7. The components and the operation determine the answer but the answer and the operation don't determine the components. Yet, any code must have this invertibility property: It must be possible to reconstruct the message from the encrypted message. This puzzle asks you to resolve this conflict: Assign numbers to get unique decodability through an operation that is not invertible.

26

The Dictator and His Mistress

He said his name was David, bothering neither to give a last name nor to state his profession. There was something about the purposefulness of the walk and precision of his speech that suggested a military connection, all of this despite his unruly blond curls. "A particularly brutal and fickle general named Slit runs the Libdan army in its war against the rebels," he explained. "He is quite fearful of being shot by his enemies, however, so he travels with one unit of soldiers and then sends his orders out by motorcycle messenger to his top commanders. On these roads, the motorcycles take almost an hour to arrive. The commanders relay messages to their subordinates in the same way.

"From what we can gather, Slit's army works according to strict hierarchical principles. Each subordinate will do whatever his superior last commanded on pain of death and will pass on orders to his subordinates. There are two commands: attack or retreat. Our goal is to find the unit commanded by Slit and possible hierarchies based on the patterns of command decisions that our satellites see among the fighting units associated with each officer. By the way, in the absence of orders from a superior, a unit will do what it wants but will nevertheless send messages to its subordinates—petty tyrannies abound.

"Two days ago there was a 16-hour battle involving 12 Libdan army units, one attached to the general, and one attached to every officer in the hierarchy. Taking 1 to be attack and 0 to be retreat, here is what the units did in the first hour: 101110011001. That is, unit 1 attacked; unit 2 retreated; units 3, 4, and 5 attacked, and so on. Here are all the configurations in all the battles listed in order by hour. Our informants have suggested that unit 1 commands nobody else.

101110011001,
011101111111,
100110101111,
110111011110,
011001010000,
001000100001,
100110101111,
111111011110,
011001110001,
100110101111,
110111011110,
010001010000,
001000000000,
000000100001,
100110001110,
011001010000.

"Remember that it takes an hour for a command to travel from a commander to a subordinate. You can assume therefore that the subordinate will do what the commander did the hour before. You can also assume that each commander has at least two subordinates."

"Hold on. Could we try a little example?" 13-year-old Liane asked, putting down her electric guitar. "Suppose there were just three units (the general and two subordinates) and we observed the following attack configurations:

010,
101,
000.

Cybernovice: What would you then suspect?

Then we would strongly suspect that the general commanded the middle unit (unit 2) directly, because the other units always copy what the middle unit does with an hour's delay. Is that correct?"

"Precisely," said David. "Do you have security clearance . . . Sorry, forget it. Can you discover the hierarchy and which unit the general was with?"

Liane went back to her guitar but stared at the sheet of paper that David put out. Then she put her guitar back on its stand.

1. Advanced cybernovice: Please give it a try. Liane and Dr. Ecco worked on the problem and arrived at a hierarchy consisting of a four-level unbalanced tree.

David reviewed their analysis. "This makes a lot of sense. I'll take it back to my colleagues, but let me continue my story. It turns out that the day did not go well for General Slit. He decided, therefore, to bring in his mistress Silky. She became a commander though we don't know where in the hierarchy. We also suspect he changed the hierarchy. It seems that Silky has a mind of her own, however, and started making decisions in the 16-hour battle that occurred 3 days later. Can you figure out the units Slit and Silky commanded directly as well as the new hierarchy? Continue to assume that each commander has at least two subordinates."

001010010111,
110011011001,
111111101111,
111100101111,
000000010000,
000010010000,
110011011001,
111111111111,
111111101111,
111101111111,
001111100110,
111101101111,
001100110110,
000011010000,

111111111111,
111111101111.

2. Cyberexpert: Liane and Dr. Ecco were able to find an answer consistent with the data in which we have reason to believe that Silky disobeyed exactly three times and that the hierarchy has only three levels. Would you like to give this question a try?

Solutions

The general method for solving such a problem is to consider each pair of columns in the attack/retreat matrix and see whether either member of the pair commands the other. The unit associated with column i commands the unit of column j if j at row $k+1$ does what i did at row k (one above). Once you find these relationships, you can then build a hierarchy (or perhaps several). Then you check other constraints.

1. The general is with unit 3. The commander of unit 3 sends orders to the commanders of units 7 and 12. The commander of unit 12 sends orders to 4, 9, 10, and 11. The commander of unit 7 sends orders to 1 and 5. The commander of unit 5 sends orders to 2, 6, and 8. Here are the hierarchies in outline format:

```
3
 7
   1
   5
     2
     6
     8
 12
   4
   9
   10
   11
```

2. When the mistress is involved, the general is associated with unit 8, the mistress with unit 6. The general sends orders to the commanders of 5 and 6. The commander of 5 sends orders to the commanders of 1, 2, 9, and 12. The commander of 6 (Silky) sends orders to 3, 4, 7, 10, and 11. She disobeys the general three times. Here are the hierarchies in outline format:

```
8
  5
    1
    2
    9
    12
  6
    3
    4
    7
    10
    11
```

There may be other possibilities. Can you find them?

Inspiration and Offshoots

Traffic analysis is that branch of cryptography that consists of looking at the pattern rather than at the contents of the messages. During World War II, the British used the flurry of German naval messages in the North Sea to predict an attack. A few years later, Soviet scientists inferred that the Americans were working on an atomic bomb thanks to the sudden disappearance of articles on the subject in U.S. science journals.

In this puzzle, the "traffic" consists of the attacks and retreats. The problem is to infer leadership. For the first part of the puzzle, the basic method is to detect that position *i* could be the direct commander of position *j* if every value at position *i* were followed 1 hour later by the same value in position j. In the second part of the puzzle, the potential duplicity of Silky made this a frequency determination: i leads j with sufficient frequency if j copies i with 1 hour's delay most of the time. As the solutions show, several different leadership patterns are possible.

27

Untangling the Rosetta Web

"If our friends at the FBI are telling the truth, this group is sending terrorism instructions over the Web," the woman said. Our visitor had the inquisitive eyes of a mathematician, mixed with the reticence of a National Security Agency officer. She was also Ecco's longtime friend Karmen Simon. Ecco knew too well that he shouldn't ask questions. Whereas he had had trouble with the NSA in the past (see the book *Codes, Puzzles, and Conspiracy*) his friendship for Karmen overwhelmed his antagonism to the institution.

"They are not very sophisticated," Simon continued. "A cursory analysis shows a single substitution code. But the problem is that the message is spread on a bunch of Web pages. These have titles coming from dried fruits: raisin, date, fig, prune, apricot, pineapple, grapefruit, currant, and coconut. We believe the recipient is meant to lay out the pages in a specific order. The first question is which order?

"Our FBI friends tell us that this group, calling themselves 'Remember Waco'—a group of antifederalists—normally works from a template and we want to exploit that fact."

"What do you mean by a template?" Liane asked.

"Well, we have a kind of hyper-Rosetta stone," Simon responded.

"We know that the graph has nine pages numbered consecutively: 1, 2, 3, 4, 5, 6, 7, 8, 9. The message is laid out in numerical order on those nine pages. According to our informants, the hyperlinks are supposed to link the pages as follows:

2 → 6
1 → 5
4 → 4
2 → 2
5 → 7
4 → 8
8 → 9
4 → 7
3 → 6
8 → 3
8 → 4
1 → 4
9 → 7
8 → 9
5 → 2
6 → 2
5 → 7
6 → 2
8 → 5
7 → 6

Each row represents a hyperlink. So, for example, page 2 has a hyperlink to itself and to page 6. At the same time, page 6 has two hyperlinks to page 2 and page 5 has one. That's the classic arrangment.

"The hyperlinks we actually see, however, come in an entirely different form:

fig → prune
grapefruit → grapefruit
apricot → currant
pineapple → apricot
raisin → currant
grapefruit → currant
apricot → pineapple
pineapple → grapefruit
pineapple → date
fig → prune
pineapple → raisin
coconut → grapefruit
raisin → currant
raisin → coconut
prune → fig
date → fig
raisin → prune
grapefruit → pineapple
prune → prune
currant → fig

So, the first problem is to determine which dried fruit corresponds to page 1, which to page 2, etc. All we know for sure is that the edges should correspond at least approximately, so if apricot is page 1, then apricot should point to two pages, corresponding to pages 4 and 5. What confuses things is that according to our informants two hyperlinks may have been reversed. We want to identify those two if possible. The second problem is to decode the message. Some think it's better to start with the decoding, under the theory that the decoding will help the ordering. I don't know. Since they change the messages more often than the links, we are certainly interested in the ordering in any case.

"Here is the text associated with each page:

apricot: "rrmlbgvp";
coconut: "6nlm4bgu";
currant: "rmar6mvx";
date: "8nb8vmhw";
fig: "xu6gllm7";
grapefruit: "wdddm3r8";
pineapple: "un6mx6m6";
prune: "rmx5mram";
raisin: "a7m9rgm9".

"Assuming you can order the pages, the message in cyphertext is just the concatenation of the messages in each page. As I mentioned, we know from our agents that the code is a single substitution code in English mapping from all lowercase letters, all digits, and the space character to the same alphabet, viz., all lowercase letters, all digits, and the space character. In this case, however, the encrypted text has no space character. That may or may not mean that the clear text has no blank spaces.

"The basic problem as I see it, Dr. Ecco, is approximate graph isomorphism," I volunteered. "It's perhaps made easier by the fact that some edges are represented multiple times. But the basic principle is the same. We want a mapping from fruit to page numbers and then we want to do a decoding."

"Quite right, Professor," Ecco said to me. "I understand that your field hasn't quite decided how difficult even exact graph isomorphism is."

"True," I admitted. "No efficient good algorithms are known, but the problem is not known to be hard, either."

Cybernovice: Your job is to solve these three problems: Find the correct ordering among the pages, identify the reversed edges if any, and then decrypt the message as much as possible. (Hint for later: There are three 2s, two xs and a v in the license plate.)

Ecco and Liane worked on the problem for many minutes. As he handed the written solution to Simon, Ecco smiled and said, "Personally, I like dates better than figs."

Solution

The first problem is to find a mapping from dried fruit to page numbers. Here is what Dr. Ecco found:

coconut → 1
prune → 2
date → 3
grapefruit → 4
raisin → 5
fig → 6
currant → 7
pineapple → 8
apricot → 9

There were two reversed edges as Simon had feared. They are (i) apricot → pineapple, which should have been pineapple → apricot, and (ii) raisin → coconut, which should have been coconut → raisin.

When we lay out the text in order of page number and concatenate, we get

"6nlm4bgurmx5mram8nb8vmhwwdddm3r8a7m9rgm9xu6gllm7rm
ar6mvxun6mx6m6rrmlbgvp"

The corresponding cleartext is "the cargo is on uhaul vxx222 bound for figtree do not light it too early."

Inspiration and Offshoots

The Web presents many opportunities for keeping secrets. Pages are linked but unordered. Text can be hidden in pictures or even in music. This puzzle forces the would-be decryptor to find the order among pages and then to decrypt some text. Some practically unbreakable codes may be built from these ideas, I think.

28

The Richest of Them All

"Rich people aren't supposed to talk about money," the impeccably dressed lawyer said with a smile, "but they would like to. Like many people, they want to know how they rank in the pantheon of personal fortune."

The lawyer had announced himself with a business card that read: "Blake Virginia Barrey, chief attorney, the Hectomillionaire's Club." None of us had heard of this club, but that should surprise nobody.

Barrey gave us some background: "To enter this club, you must show a bank or stock account statement with $100 million in it. Then you must be recommended by a committee of members. If you are worthy, you are chosen. Membership offers you visiting privileges to the most exclusive yacht, golf, and tennis clubs around the world, as well as priority for the penthouse suites of the world's greatest resort and metropolitan hotels from Zimbabwe to Zurich.

"Each year there is an annual meeting on an island in the Caribbean owned by the club. There, the hectomillionaires meet to discuss new business ventures, many involving exclusive monopolies from mineral-rich nation states."

"The few get rich and the many walk barefoot in the mud," Ecco observed, his eyes narrowing.

"It's very complicated," Barrey replied quickly in his cheery voice, aiming for a lawyerly appeasement. "Some of these people are good and some are less good. What I'm asking you for today is help with a problem that has nothing to do with politics, merely with money."

Ecco scowled, but 11-year-old Liane was already sitting on the edge of her seat preparing for the puzzle.

Barrey continued: "We have 20 people around the world. They are all hectomillionaires ($100 million at least), but none have more than $10 billion. They are rich enough, however, that they all round their wealth to the nearest $100 million. Several may have the same number of hectomillions. They want to know how many hectomillions the richest and poorest among the 20 have and how big those rich and poor subgroups are. Nobody should learn who else is in any subgroup, unless all or all but one are in some subgroup or there is large-scale collusion. You can assume that they will be honest in the information they give and in any calculations they are asked to do."

A few seconds' pause brought Liane's first question. "Since they care about accuracy only to the nearest $100 million and since nobody is richer than $10 billion, there are only 100 possibilities. We can use the same logic here as we would with 20 people whose wealth fell between $1 and $100, where everyone estimated his or her wealth to the nearest dollar. Right?"

"True," said Barrey. "But then I wouldn't be their lawyer."

We all laughed. Even Ecco managed a smile.

"Let's see," said Liane. "If they all were in one room, then I could think of a fairly easy solution if you gave me 100 mailboxes."

Cybernovice/Cyberexpert: Do you have any ideas?

"What would you do then, young lady?" Barrey asked.

"Well, I'd start by labeling each mailbox with a hectomillion (1 hectomillion, 2 hectomillions, . . . , 100 hectomillions) and would lay them out in that order," Liane responded. "I'd give each hectomillionaire 100 identical envelopes, and 99 pieces of paper saying no and 1 piece saying yes. Each hectomillionaire would go to a private room where he or she would place the yes in one of the envelopes and the nos in the other envelopes. Then the hectomillionaire would place one envelope in each mailbox, putting the one with a yes in the mailbox corresponding to his or her wealth."

"Nice work, Liane," I said.

"Thank you, Professor Scarlet," she replied, "but this does not help when they must communicate by phone and the phones may be able to trace back calls."

"A promising young cryptographer named Camus Tiger once mentioned a trick," Ecco said. "It goes like this: If I am thinking of a number N that is known to be less than another number M, then I can give Alice, Bob, and Carol three numbers N_1, N_2, and N_3 respectively (all smaller than M) such that $(N_1 + N_2 + N_3) \bmod M = N$. Knowing even two of those numbers, nobody can infer my N. Maybe one could use that."

I looked at Ecco and knew that he had a solution in mind. But he said no more. He picked up a journal of genetics and started to read it. Barrey sat down with a sigh and stared up at the ceiling. Only Liane was active, scribbling on the paper, drawing tables of numbers.

Suddenly, she stopped, looked at Barrey, and said, "I've got it. I can find the wealth of the poorest and the richest subgroups and how many are in each of those subgroups."

Liane explained her technique, taking full advantage of Camus Tiger's trick. Her method not only revealed the wealth of the poorest and richest hectomillionaires, but also the sizes of the poorest and richest subgroups. That was what Barrey wanted. Her method leaked a bit more information as well: the number and wealth of all intermediate subgroups. Still, the method divulged no information identifying any individual.

Barrey listened to her explanation twice. He took notes then showed them to Liane, who corrected them. He reconfirmed several times that (i) no outside person was involved (they are very private) and (ii) the phone calls in her protocol did not have to be anonymous (the callee would know the caller, otherwise pranksters might enter in). He said she could assume that the phone calls would be private (no tapping and no conference calls). He reread them and then claimed that he understood. He reached into his pocket and pulled out a certified check and handed it to Ecco. Ecco put it on the table and led the visitor to the door.

After Barrey left, Ecco showed the check to Liane. It was for $300,000. "We'll put it in your college fund," he said. "At least some good will come out of these financial raptors."

Cybernovice: Liane's solution requires that all 20 people communicate with each other at least once, requiring nearly 190 phone calls. With her method, 19 people must collude to infer the 20th person's wealth. Try to find a method that finds the wealth of the richest and poorest subgroups, offers the same collusion protection as Liane's method (i.e., at least 19 people must collude to infer the 20th's wealth), and can be done with as few phone calls as possible.

Solution

The solution makes use of modular arithmetic. Let me give you some examples: $4 \bmod 7 = 4$, $10 \bmod 7 = 3$, $51 \bmod 7 = 2$. That is, $x \bmod 7$ is the remainder that you get after dividing x by 7. Our protocol uses mod 31. Any number over 20 would do.

Let's simplify the protocol to four people: Alice, Bob, Carol, and David. Suppose that Alice has $1 billion, Bob has $2 billion, and Carol and David each have $3 billion. Each of them maintains 100 buckets. Bucket 10 matches $1 billion, bucket 20 matches $2 billion, and bucket 30 matches $3 billion. Now Alice will send to everyone, including herself, a number corresponding to each bucket. Simplifying to make the numbers easier to understand, she sends 4 to Bob, 6 to Carol, 10 to David, and 11 to Alice (herself) for all buckets except for bucket 10, for which she sends 4 to Bob, 6 to Carol, 10 to David, and 12 to herself. Notice that $(4+6+10+11) \bmod 31 = 0$; however, $(4+6+10+12) \bmod 31 = 1$.

Bob also uses this pattern except that he sends 4 to Carol, 6 to David, 10 to Alice, and 11 to himself for every bucket except 20. For bucket 20, he sends 4 to Carol, 6 to David, 10 to Alice, and 12 to himself.

Carol sends 4 to David, 6 to Alice, 10 to Bob, and 11 to herself for every bucket except 30. For bucket 30, she sends 4 to David, 6 to Alice, 10 to Bob, and 12 to herself.

David sends 4 to Alice, 6 to Bob, 10 to Carol, and 11 to himself for every bucket except 30. For bucket 30, he sends 4 to Alice, 6 to Bob, 10 to Carol, and 12 to himself.

Now, each person calls Alice to give the totals of each bucket. For all buckets except 10, 20, and 30, Alice will receive 31, 31, 31, 31. Summing this mod 31, she gets 0. For buckets 10 and 20, she will receive 32, 31, 31, 31. Summing this mod 31, she gets 1. For bucket 30, she will receive 32, 32, 31, 31. Summing this mod 31, she gets 2. Thus, Alice will be able to count those that have each amount of wealth. The one problem with this simplified solution is that everyone follows the same pattern, so Alice can in fact tell who has each amount of money by who is reporting 32 for a particular bucket.

The real solution is the same except that randomness is used to protect information. Imagine again that each person maintains 100 buckets, one for each number of hectomillions between $100 million and $10 billion. He or she will receive numbers from other people and put them into buckets. Here is how. Each person gives 100 numbers (randomly generated) between 0 and 20 inclusive to each other person. The values $n[1,i,j], n[2,i,j], \ldots, n[100,i,j]$ are the 100 numbers i gives to j. $n[1,i,j]$ goes into bucket 1 at j's site, $n[2,i,j]$ goes into bucket 2 at j's site, and so on. Each person receives 100 numbers from himself, too. If person i has M hectomillion dollars, he will make the sum over j of $n[M,i,j]$ be $1 \bmod 31$. Thus the contribution that i makes to bucket M at all sites sums to $1 \bmod 31$. For all L unequal to M, i will make the sum over j of $n[L,i,j]$ be $0 \bmod 31$. Each person j will then sum the numbers in bucket 1, sum the numbers in bucket 2, etc., and then take the modulus of 31. A further sum will be made for all the totals of bucket 1 at all sites and again a modulus of 31 will be applied to the result. Similarly for buckets 2 through 100. This second sum can all be done at a single site, assuming honesty.

Imagine that the total for a bucket q is 1. That means that exactly one person has q hectomillion dollars, since all other contributions will cancel out. If the total is 8, say, then eight people have q hectomillion dollars.

There is no possibility of collusion unless everyone gangs up on you, because that is what it takes for them to infer the contribution you have made to each bucket.

Inspiration and Offshoots

The general setup of this solution comes from Doug Tygar, a professor at Berkeley as of this writing. He gives a great talk about auctions in which he shows how an auction can be implemented with the efficiency and privacy of a sealed bid but the effect of a conventional (English) auction. Dr. Ecco used similar mathematics to solve this problem. The problem with these solutions is that they assume that all the calculations done by each millionaire will be correct. What if at most two millionaires might lie?

Part VII

Pattern Mathematics

29

Anthrax and Dr. Windswift

Dr. Roger Windswift of the Centers for Disease Control and Prevention had chosen his career of ethnobiologist well. Kayaker, underwater photographer, naturalist, entirely comfortable in the jungle or makeshift lab, he seemed confined by Ecco's MacDougal Street apartment. After shaking our hands, he paced the room as he described his problem, touching objects to sense their texture without interrupting his train of thought. His story unfolded in academic innocence.

"In the mid-1980s, Jeffrey Palmer and L. A. Herbon showed that cabbages and turnips, while close in DNA sequence, differ from each other mostly through a series of reversals," he said. I saw Ecco's face settle into an expression of bemused indifference.

"A reversal of a string between positions x and y," Dr. Windswift continued, "is a permutation of the string in which the subsequence between x and y is reversed but nothing else is changed. So, starting with a string whose indexes are numbered from 0 to n:

$$0123 \cdots (x)(x+1) \cdots (y-1)(y)(y+1) \cdots n,$$

This adventure occurred in October 1998.

he obtains a new string with a reversal in the middle:

$$0123 \cdots (y)(y-1) \cdots (x+1)(x)(y+1) \cdots n.$$

We denote such permutations by the low and high indexes of the reversed subsequence, in the above case (x, y). For example, given the string abcdefg the reversal (2, 4) yields abedcfg [leave ab alone, reverse cde, and leave fg alone].

"A strain of bacteria closely related to anthrax has recently appeared. We suspect that it was manufactured." He stopped his pace for a moment and swept his eyes to each of us.

Dr. Windswift went on. "Fortunately, this particular strain does not appear to be contagious, but we want to be ready for the next one. Our strategy will be to create antidotes by doing reversals of some secret bacteria that we have. This will serve as a kind of vaccine.

"We have recently developed a protocol that that will transform a bacteria population A into a new one B that is identical to A except for one reversal that we can control. The protocol takes a day to complete. If we want to do five reversals, then, we will need 5 days."

"Is there no way to work on many reversals at the same time?" Liane asked.

Windswift looked at the 10-year-old girl for a moment, then said, "Good question. As far as we can tell, the answer is no. That's why we're here. We need to create these antidotes as fast as possible. That means we have to figure out the fewest number of reversals to transform some critical protein sequence to another one, all *in vivo*."

"Could you give us an example of a reversal problem?" Liane asked.

"Sure, I'll give one that I had trouble doing efficiently. Find the smallest number of reversals that take abcdefg to cfegbad. I challenge you to do it in three reversals," Dr. Windswift answered.

Cybernovice: Please give this a try.

Liane thought for a minute and then wrote down:

(0, 2), producing cbadefg;

(1, 5), producing cfedabg;

(3, 6), producing cfegbad.

"Nice job, Liane," Ecco said chuckling. "Well, Dr. Windswift, you can see we have the talent to do rapid, directed evolution. Please give us a real problem now."

"Very well," Dr. Windswift said. "But please understand that the sequence I'm concerned with could damage lung tissue. Please be careful not to reveal this information."

"Don't worry," Ecco said with a smile. "We don't even know which sequence of nucleotides each of your symbols represents."

"Here is the start sequence: mhtvkllcvvfsclcavawasshrqpchsp. We want to produce the following sequence through as few reversals as possible: awavchtfrhsscpchvmsvkllvqlcasp. Can you do it in under 14 reversals? Two weeks is about all the time we have, but we can do 14 reversals in 2 weeks."

Ecco and Liane worked on the problem for a while. Liane came up with a solution first.

"I can do a few better than you asked for, Dr. Windswift. I've written it down in steps so you can follow it easily. Along with each step, I write the low and high indexes of the subsequence being reversed."

> 1. *Advanced cybernovice: Can you do better than 14? How small a number of reversals can you find? Nine is the best I know of.*

"Wow," Dr. Windswift said. "Do you have any time after school to do some consulting?"

"My uncle keeps me busy enough, thanks," Liane answered sweetly.

Dr. Windswift left, but returned a month later. "The anthrax manufacturers have found a new mutation path. Not only can they cause reversals in subsequences, but they can cause subsequences to rotate by one to the right. For example, given the sequence abcdefg, the rotation (1, 5) would yield afbcdeg. [The f moves to position 1 and bcde are moved one to the right; remember that the sequence numbering starts at 0]. Fortunately, we've found a process to do a rotation in a day, but now our mathematical problem is to figure out the shortest possible combination of rotations and reversals to transform one sequence to another. This seems hard to us."

"Could be. Please let's have a try," Ecco said.

"Here are the sequences," Dr. Windswift said. "Start again with mhtvkllcvvfsclcavawasshrqpchsp but end this time with mhtcskaavv-walcsfvqpchspshrvcll."

2. Cyberexpert: Try to find a short sequence of rotations and reversals to get the ending string from the starting string. Remember that the rotations take a subsequence and rotate it by one position to the right. Use at most seven rotations and reversals.

Ecco worked on the problem for several minutes in silence. Then he raised his head and asked: "Is there any reason to believe that it is easier to rotate just after reversing?"

"Not that I know of," Dr. Windswift said. "Why do you ask?"

"No reason. I just suggest you discuss this possibility with your chemists," Ecco said, as he handed Dr. Windswift an eight-step transformation.

Two months later, a small item in a newspaper reported a disease outbreak following a pistachio party in Los Angeles. An unidentified scientist from the Centers for Disease Control was on the scene. A few days later, health officials issued an advisory that certain kinds of pistachios should be avoided and a second advisory that people with rashes behind their ears should call the CDC in Atlanta. Then the news media returned to its obsession with the stock market and sex scandals.

Solutions

1. The first problem demanded a solution using reversals only. The best answers I received required nine steps. The first such solution came from Eric Haines, who proposed the following:

STEPS	REVERSALS	TRANSFORMATION
start		mhtvkllcvvfsclcavawasshrqpchsp
1	(12, 19)	mhtvkllcvvfsawavaclcsshrqpchsp
2	(9, 14)	mhtvkllcvawasfvvaclcsshrqpchsp
3	(7, 14)	mhtvkllvfsawavcvaclcsshrqpchsp
4	(3, 8)	mhtfvllkvsawavcvaclcsshrqpchsp

STEPS	REVERSALS	TRANSFORMATION
5	(4, 23)	mhtfrhssclcavcvawasvkllvqpchsp
6	(1, 12)	mvaclcsshrfthcvawasvkllvqpchsp
7	(5, 24)	mvaclqvllkvsawavchtfrhsscpchsp
8	(2, 27)	mvhcpcsshrfthcvawasvkllvqlcasp
9	(0, 17)	awavchtfrhsscpchvmsvkllvqlcasp

Eric also pointed out that this puzzle is similar to those in the games section of *What To Do After You Hit Return* published in 1975 by the People's Computer Company. The pages he sent me talked about reversals that started from the beginning of the string, which is a little easier than these general reversals.

2. Dr. Burghart Hoffrichter was the first to present a seven-step solution for the second problem and to point out that rotations didn't help in that case:

STEPS	REVERSALS	TRANSFORMATION
start		mhtvkllcvvfsclcavawasshrqpchsp
1	(5, 29)	mhtvkpshcpqrhssawavaclcsfvvcll
2	(11, 25)	mhtvkpshcpqvfsclcavawasshrvcll
3	(5, 22)	mhtvksawavaclcsfvqpchspshrvcll
4	(3, 11)	mhtcavawaskvlcsfvqpchspshrvcll
5	(6, 11)	mhtcavvksawalcsfvqpchspshrvcll
6	(4, 8)	mhtcskvvaawalcsfvqpchspshrvcll
7	(6, 9)	mhtcskaavvwalcsfvqpchspshrvcll

Inspiration and Offshoots

Genetic evolution remains a great mystery. A relative handful of genes (a few tens of thousands in humans) determines so many attributes of an organism with remarkable precision. For example, a single nucleotide mutation in one gene can lead to a blood disease. Another can be equally precise yet comical, like the still-to-be-discovered mutation that determines which fingers a child will suck. The remarkable power of evolution is that mutations can occur relatively easily at the DNA level whereas at the phenomenological level they seem so enormous. It's as if a single nail struck at the foundation of a building could lead to a skyscraper rather than a cottage. A mildly famous college dropout named Bill Gates co-wrote a seminal paper on the reversal problem: William H. Gates and Christos H. Papadimitriou, "Bounds for Sorting by Prefix Reversal," *Discrete Mathematics* 27 (1979), pp. 47–57.

30

The Virus from the Spy

He meant to look like a professorial colleague, but something about his sense of urgency suggested spy. "We have to find the sequence of this microvirus extremely fast, Dr. Ecco," the bearded biochemist told us, never bothering to adjust the asymmetric placement of the glasses on his face. He introduced himself as Bill Smith and said he worked for "the government."

"I thought there were machines for that," Ecco replied.

"Well, yes, but this sequence comes to us from an agent in a certain, well, not-so-friendly developing country," Smith replied. "We have only the data I'm going to give you, not the sequence itself. This virus can kill and appears to be contagious, partly because of its small size. Our agent reports many deaths at the factory producing the virus. If released, it could cause untold more deaths."

"Sounds bad. What do you know?" Ecco asked.

"Well, we know about two of its sequences," said Smith, clearly in a rush. "To analyze them, our agent used three restriction enzymes. He then separated them using agarose gel electrophoresis. . . ."

This adventure occurred in September 2000.

Smith paused and looked at our blank faces. Liane had taken to cutting a spiral in a returned homework paper.

"I see that you're not following," he said. "Let me start more slowly. It's just that there is so little time. You know that DNA can be viewed as a sequence of As, Cs, Ts, and Gs. It exists most commonly in the form of a double helix, but because one side of the helix determines the order of the other side, we will pay attention only to one side and pretend we have just a single strand consisting of half of the helix. The sequence is read left to right from what is known as the 5′ [read "5 prime"] end towards the 3′ end.

"Certain kinds of proteins called restriction enzymes cut the sequence at particular points. In our case, we have three, which we'll call r1, r2, and r3. Restriction enzyme r1 will cut a sequence at CT and split C from T. Similarly, r2 will cut at AG and r3 will cut at TT. If there are three in a row, it will cut only once, however. After one or more restriction enzymes performs its cuts, the ministrands that result can be ordered by size. When they are small enough, the ministrands can be sequenced directly. Given the sizes after each cut and the sequences of the final ministrands, we must figure out the sequence of the whole strand. In particular:

- if r3 sees TTT, then it will cut this as T TT.
- If r3 sees TTTT, then it will cut this as T TT T.
- If r3 sees TTTTT, then it will cut this as T TT TT.

"Let me give you an example. Suppose we have the sequence CATTCTGTATA. If we cut it with r1 alone, we will split the sequence into CATTC and TGTATA of sizes 5 and 6. If we cut with r2 alone, no split will occur. If we cut with r3 alone, we will get CAT and TCTGTATA, of sizes 3 and 8. If we cut with both r1 and r3, we will get CAT, TC, TGTATA.

"Of course, if the agent's data told us the sequences and their orders, our problem would be over. Unfortunately, all we know is their sizes and the contents of the final arrangement. So, revisiting the example, imagine that I told you only the following: (i) Splitting with r1 gives sizes 5, 6; (ii) splitting with r2 gives no split; (iii) splitting with r3 gives sizes 3 and 8; (iv) splitting with r1, r2, and r3 gives three cuts [which I'm deliberately writing in order of size, because that's the order in which all these results are given]: TC, CAT, TGTATA. What would you do?"

Cybernovice: Before reading on, see if you can figure out the ordering.

Liane had stopped cutting spirals and was listening intently.

Smith continued, "To infer the order, note that none of our enzymes cuts at AT or AC or CC junctions. So TC can bind only to TGTATA [which is also consistent with the fact that r3 splits the sequence into lengths 3 and 8]. Further, TGTATA must be on the right because it cannot bind to anything. This gives us a single possible ordering—CAT TC TGTATA."

"OK, that's clear," 12-year-old Liane said, eyes sparkling. "Tell us what we need to know for your super-dangerous microvirus."

"Here is the data about the smaller sequence," Smith said. "The restriction enzymes cut as stated before [r1 cuts at CT, r2 cuts at AG, and r3 cuts at TT]:

CUTS	SIZES OF CUTS (IN ORDER OF SIZE)
r1	2, 2, 2, 7, 9, 11, 17
r2	2, 4, 4, 11, 29
r3	1, 49
r1, r2	2, 2, 2, 2, 2, 2, 4, 7, 7, 9, 11
r1, r3	1, 2, 2, 2, 6, 9, 11, 17
r2, r3	1, 2, 4, 4, 10, 29

"Finally, the last thing we know is the collection of sequences in order of size after being cut by all three restriction enzymes:

T
TC
TC
TC
TA
GA

GC
GATA
TCCATC
GTCCGTC
TCACACGGC
TCGCACACGGA

"Our question to you now is what is the entire sequence in its natural order? If you can't be completely precise, we understand, but the more you can tell us, the easier it will be for us to find an antibody."

"What do the repeats, like the ones for TC, mean?" Liane asked.

"According to our agent, they mean that TC appears three separate times in the sequence," Smith replied.

1. Advanced cybernovice: Give it a try. Liane started out using scissors to cut out the subsequences that would be cut by all three restriction enzymes.

"That wasn't too hard," Liane said after 20 minutes, handing her solution to Smith. "Give us the other."

"I had heard that you were clever," Smith said after verifying the result. "Now I know it's true. The second sequence is quite a bit longer:

CUTS	SIZES OF CUTS
r1	6, 12, 21, 29, 39
r2	3, 4, 6, 11, 11, 12, 13, 19, 28
r3	2, 5, 6, 6, 7, 10, 13, 16, 19, 23
r1, r2	2, 2, 3, 4, 5, 6, 7, 11, 11, 12, 12, 13, 19
r1, r3	1, 1, 2, 5, 5, 6, 6, 7, 9, 9, 13, 14, 14, 15
r2, r3	2, 2, 2, 2, 3, 3, 3, 4, 4, 6, 6, 7, 8, 8, 10, 11, 12, 14

"Results of all three restriction enzymes give cuts that look like this:

T
T
GT

GT
GT
GC
TA
TT
GGA
GGT
TTA
TACA
TGGAA
TGGTGT
GGCGGA
GGACGTC
TCGTATA
TATGCGAA
TTACATGTC
TATCGCGAC
TACGGCCCCGA
GCCGCGATCCAT

2. *Cyberexpert: Try to find this longer sequence. If there are ambiguities in the order, say so.*

Solutions

Generate a set of sizes and their sequences for each kind of restriction and each combination of restrictions. From this, figure out the entire sequence. Here is an intermediate result:

0: T;
1–6: TCCATC;
7–8: TC;
9–10: TA;
11–12: GA;
13–19: GTCCGTC;
20–28: TCACACGGC;

29–30: TC;

31–41: TCGCACACGGA;

42–45: GATA;

46: GC;

48: TC.

Here is the whole sequence: TTCCATCTCT AGAGTCCGTC TCA-CACGGCT CTCGCACACG GAGATAGCTC.

The longer sequence is: GGAGTTACGG CCCCGAGTTG GTGT-TACAGT TTACATGTCT TTTATCGCGA CTGGAAGGCG GAGGT-TATGC GAAGGACGTC TTTAGCTAGC CGCGATCCAT TCGTATA.

Inspiration and Offshoots

Determining a full sequence by cutting it into small pieces flies in the face of all macroscopic intuition. For example, no Egyptologist in Napoleon's time would have requested Napoleon's soldiers to break apart the Rosetta stone with a sledgehammer to make for easier reading. Future technologies may well render the DNA-cutting approach obsolete by permitting researchers to read long sequences directly (three colleagues of mine are working exactly on such an approach). In the meantime, however, the high precision of the restriction enzymes offers a kind of reconstructive glue, allowing the restriction enzyme–based "shotgun sequencing approach" to construct an approximation to the human genome. It's as if Napoleon's soldiers had found the Rosetta stone in fragments, but scholars could reconstruct the entire tablet by matching cut edges. Still this approach requires an initial sequence. For *Jurassic Park* to be realizable, it should be possible to work with piecemeal fragments of DNA. OK, maybe I haven't drawn the right moral lessons from the movie.

31

Proteins of the Worm

She introduced herself as Elena. Looking younger than her 27 years, she had been a child prodigy, one of only two women physics students at the Moscow Physics Institute and then a researcher at the Atomic Energy Institute before converting to biology and emigrating to the United States. "The Atomic Energy Institute has most of its labs in Moscow, nuclear reactors and all. It's mostly for weapons research. I can't even tell you what I worked on. It was sensitive enough that the government in Russia didn't want to let me leave the country. But here I am. There are all kinds of ways to do things in Russia," she said with a dreamy smile I could not read.

Suddenly all business, she continued, "I am here to ask for help in figuring out a protein network. Though we don't understand the genetic details of the network, we know that they involve the most important cells in our model organism, a 1 millimeter worm called the *C elegans*."

"How exciting," 12-year-old Liane said with undisguised mockery.

"I don't like them any more than you do, girl," Elena said, "but it turns out that if I understand this network, I can perhaps attack certain cancers in people."

Liane sat up, apparently decided to hear Elena out.

"A network or a pathway is a collection of causal relationships that link the production of chemicals—proteins, metabolites, whatever. If a certain critical pathway goes wrong, it can cause cancer and other diseases. So, understanding the causality in a pathway is essential to understanding what can go wrong.

"Here's how it works. Suppose there is a path of positive causality from one protein P1 to another P2, P1 itself is produced, and none of the proteins along that path have been eliminated. Then P2 will be produced also, though the amount that is produced may vary if other paths between P1 and P2 have eliminations along the way.

"For example, if we have a linear pathway A → B → C → D → E, then eliminating A will eliminate all proteins. Eliminating B will eliminate B, C, D, and E. Eliminating C will eliminate C, D, and E. I hope you see the pattern."

We nodded.

Elena went on. "Sometimes there can be branching. For example, A may branch to B and C. Then both B and C may induce D. [See Figure 1.] In that case, eliminating B partly reduces the output of D.

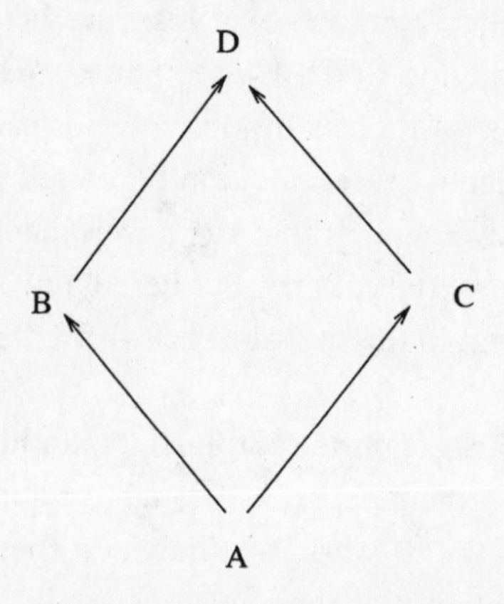

Figure 1. A network with two paths from A to D.

"The partial reduction means the circuit is analog, not digital," Ecco said with delight. "At least nature rebels against the intellectual boxcar of our age!"

"But, wait, what if you don't know the circuit?" Liane interrupted.

"That's exactly the problem," Elena replied. "We don't know the circuit. We just have certain evidence and we want to discover the most likely circuit. In my case, there are 8 proteins in the first circuit and 12 in the second. We know that all causalities are positive. In the first case, we know there is no feedback. In the second we suspect the feedback is there. Positive feedback—that is a cancer circuit."

"What data do you have?" Ecco asked.

"In each mutagenesis experiment we directly modify one of the proteins, thereby eliminating it. The others are modified through causality as explained above. There is one piece of bad news: We don't know which one we have modified, just that we have modified one."

"Here are the results of the mutagenesis experiments. Since no two of these results are the same, we know we have directly eliminated each protein in exactly one experiment and possibly others by causality. The default behavior, without mutagenesis, is that all proteins are produced at their full concentrations. Reduced concentration means that a protein is produced at a smaller concentration than normal. Zero concentration means that a protein is not produced at all [i.e., is eliminated]. So, the first line here says that A and G are produced at full concentration; D and E at reduced concentration; and B, C, F, and H at zero concentration."

Concentration Level

FULL	REDUCED	ZERO
A, G	D, E	B, C, F, H
A, B, C, D, F, G, H		E
A, B, C, E, G	D	F, H
A, B, C, E, F, G, H		D
A	B, C, D, E, F, H	G
A, B, C, D, E, G, H		F
A, B, D, E, F, G, H		C
		A, B, C, D, E, F, G, H

1. Cybernovice: All possible mutagenesis experiments are performed above, though we do not know which protein was eliminated in each case. Can you figure out at least one circuit from that information? Remember that all links are positive.

"Now for the case that may involve some feedback.

Concentration Level

FULL	REDUCED	ZERO
A, B, F, G, I, J, L	C, D, H	E, K
A, C, D, E, F, G, H, I, J, K, L		B
B, C, D, E, F, G, H, I, J, K, L		A
A, B, F, G, I, J, L	C, E, K	D, H
A, B, D, E, F, G, H, I, J, K, L		C
A, B, C, D, E, G, H, I, J, K, L		F
A, B, C, E, F, G, H, I, J, K, L		D
D, E, H, J, K	C	A, B, F, G, I, L
A, B, F, G, I, J, L	C, D, H, K	E
		A, B, C, D, E, F, G, H, I, J, K, L
A, B, C, D, E, H, I, J, K, L		F, G
A, B, D, E, F, G, H, J, K, L	C	I

2. Cyberexpert: Would you like to try to design a circuit for this? Remember that full means as much as in the case where there is no mutagenesis. Apparently the reactions don't spin off into infinite production because of some other dampening factors not involved. If you are a supercyberexpert, try to find out how many solutions there are.

After a few hours, during which time Elena clobbered me in chess three times, Ecco and Liane came up with two diagrams. After verifying that the results were consistent, Elena asked, "Are you sure no other circuits are possible?"

Ecco and Liane had to admit they weren't sure.

Cyberexpert: Can you find several?

The problem is a lot harder when repression is possible, because then the elimination of an element may actually enhance production.

Cyberexpert: Can you find a method in that case?

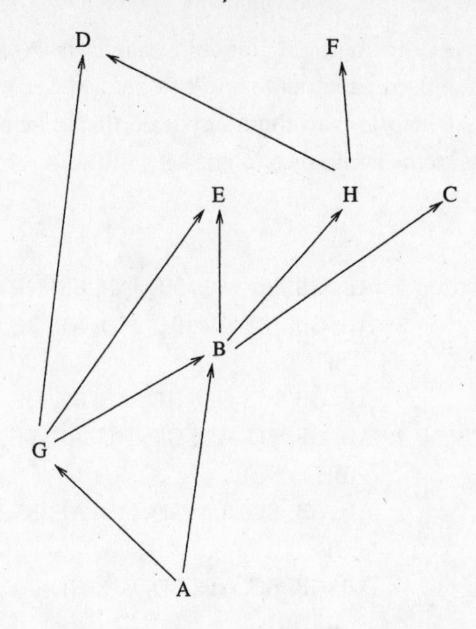

Figure 2. Solution to acyclic portion of causality puzzle.

Solutions

1. Figure 2 shows a possible solution to the acyclic circuit. In text form, the edges are: AG, GB, AB, BE, BH, GE, HD, GD, BC, and HF.

2. Figure 3 shows a solution to the cyclic circuit. In text form: JK, KE, JL, JH, HK, EH, LG, LA, LB, IC, KC, HC, GF, HD, and LI.

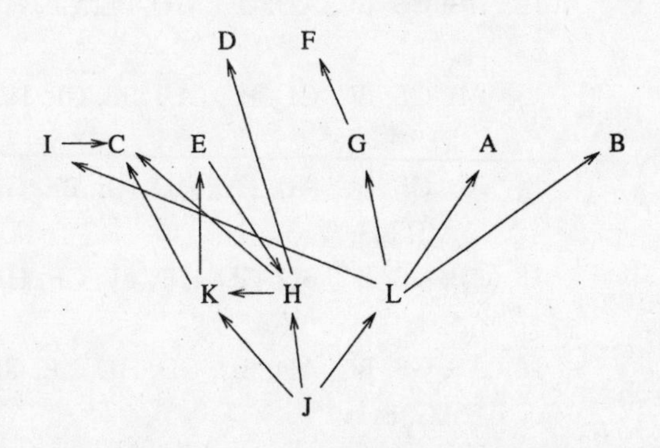

Figure 3. Circuit having feedback.

Readers used the logic programming language Prolog for this problem. Some used paper and pencil. Rodney Meyer was the first to show the 18 solutions to the first puzzle (listed below) and the 140 solutions to the cyclic second puzzle.

Protein
Permutation 1: AB, GB, BC, AD, HD, AE, BE, HF, AG, BH;
2: AB, GB, BC, AD, BD, HD, AE, BE, HF, AG, BH;
3: AB, GB, BC, GD, HD, AE, BE, HF, AG, BH;
4: AB, GB, BC, AD, GD, HD, AE, BE, HF, AG, BH;
5: AB, GB, BC, BD, GD, HD, AE, BE, HF, AG, BH;
6: AB, GB, BC, AD, BD, GD, HD, AE, BE, HF, AG, BH;
7: AB, GB, BC, AD, HD, BE, GE, HF, AG, BH;
8: AB, GB, BC, AD, HD, AE, BE, GE, HF, AG, BH;
9: AB, GB, BC, AD, BD, HD, BE, GE, HF, AG, BH;
10: AB, GB, BC, AD, BD, HD, AE, BE, GE, HF, AG, BH;
11: AB, GB, BC, GD, HD, BE, GE, HF, AG, BH;
12: AB, GB, BC, AD, GD, HD, BE, GE, HF, AG, BH;
13: AB, GB, BC, GD, HD, AE, BE, GE, HF, AG, BH;
14: AB, GB, BC, AD, GD, HD, AE, BE, GE, HF, AG, BH;
15: AB, GB, BC, BD, GD, HD, BE, GE, HF, AG, BH;
16: AB, GB, BC, AD, BD, GD, HD, BE, GE, HF, AG, BH;

17: AB, GB, BC, BD, GD, HD, AE, BE, GE, HF, AG, BH;

18: AB, GB, BC, AD, BD, GD, HD, AE, BE, GE, HF, AG, BH.

Inspiration and Offshoots

Modern science rarely involves the direct observation of the model of interest—it is not interesting science to count birds, but rather to infer the reasons for bird migrations. In biology, it is hard to observe causality in action, but one can observe the effects of disruption and thereby infer causality. This problem would be much harder under the more realistic assumption that mutagenesis could affect many proteins. How would you solve the puzzle if each mutagensis experiment affected two proteins?

32

Which Beautiful Liars to Trust?

"We're trying to crack a major drug cartel," the police commissioner explained. Commissioner Bratt was well known for his work in bringing down New York crime. It's less well known that he has been a frequent visitor to Ecco's apartment over the years. "We're using informants, but these guys are rewarded for their information, so they tend to, well, dissemble."

"That means 'lie,' doesn't it?" Liane asked with a smile.

"Well, yes, young lady," Bratt said after a brief pause, "but these are the best we could get. This cartel is one of the smoothest and smartest we've ever seen. It's run out of Argentina, with offices all over Bolivia.

"Dr. Ecco, I believe you've met the chief," Bratt continued. "The professor here talked about Solaris, also known as El Jefito because of his diminutive size, in his monograph *Codes, Puzzles, and Conspiracy*. You met him at the Casino in Punta del Este. You never figured out his role in Baskerhound's escapades, unfortunately."

"Ah yes, the Rhodes Scholar wrestler with a weakness for bimbos," Dr. Ecco said with a smile.

"Right. Especially ones wearing tight-fitting clothes," the commissioner responded. "Some of those beauties figure among our informants."

"Well, how many informants are there altogether?" Ecco asked.

"Eighteen," said the commissioner. "I think he has turned some of them, but not most I hope. The result, however, is that now we have some informants calling other ones liars. We don't know whom to trust.

"The thing is we have vetted these informants quite carefully, so we think that not more than five have turned, though it may be as many as eight," Bratt said.

"So, you know neither who lies, nor how many?" asked Liane in a slightly mocking tone.

Bratt glared at her, but nodded. "That's right. It could be any of them."

"Can you tell us who accuses whom?" Ecco asked.

Bratt laid out a list (changing the names for obvious reasons):

ACCUSER	ACCUSED	ACCUSER	ACCUSED	ACCUSER	ACCUSED
Petra	Gwenyth	Gwenyth	Mike	Olivia	Kris
Sam	Larry	Olivia	Ulm	Larry	Mike
Hillary	Petra	Jack	Larry	Tom	Hillary
Olivia	Petra	Jack	Sam	Emily	Isaac
Gwenyth	Alan	Alan	Larry	Petra	Rivera
Dave	Mike	Isaac	Mike	Mike	Larry
Isaac	Nick	Hillary	Ulm	Dave	Ulm
Hillary	Kris	Ulm	Larry	Hillary	Bob
Sam	Jack	Gwenyth	Jack	Larry	Hillary
Sam	Rivera	Gwenyth	Olivia	Gwenyth	Kris

"Well, Liane," Ecco said, "what do you make of the situation?"

"I need some more clarification," she replied. "Commissioner Bratt, am I correct in drawing the following inferences?

"Suppose X accuses Y of lying. Then

- At least one is a liar.
- If we know that X tells the truth, then Y is a liar.
- If we know that Y tells the truth, then X is a liar.
- If X lies, then his accusations may or may not be true."

"Absolutely our thinking," said the commissioner, visibly impressed. "The last point is particularly important. Lying informants may still point the finger at other liars. Also, we don't fault truth-tellers for failing to accuse liars. They simply may not know enough."

"And our job," Liane went on, "is to find the smallest set of liars who explain all these accusations?"

"Right again," Bratt said, as he glanced anxiously at Ecco.

Ecco was nibbling at a scone and staring at the list. Liane doodled while she looked at it.

After a few minutes, Ecco wrote eight names on a Post-it and handed it to the commissioner. "Here is a set of eight liars who could explain all these accusations. Liane, did I make a mistake? Can you solve the problem with fewer liars?"

"I'm not sure," Liane said. "There have to be at least six, though."

Cybernovice: First, why do there have to be at least six? Second, are eight liars a sufficient number to explain all these accusations? If so, which eight?

Cybernovice, but hard: Could there be fewer than eight liars? If so, who would they be?

Solutions

The following is a non-overlapping set of accusations. Since each accusation must include at least one liar and no two of these have the same individuals, there have to be at least six liars:

ACCUSER	ACCUSED
Petra	Gwenyth
Sam	Larry
Dave	Mike
Isaac	Nick
Hillary	Kris
Olivia	Ulm

Here is the solution with eight (there may be others; can you find them?): Isaac, Hillary, Larry, Olivia, Gwenyth, Dave, Sam, and Petra.

In fact, there must be at least eight, because either the accuser or the accused must be lying in a pair and there are the following nonoverlapping pairs: Gwenyth, Alan; Dave, Mike; Sam, Jack; Ulm, Larry; Olivia, Kris; Tom, Hillary; Emily, Isaac; Petra, Rivera.

Inspiration and Offshoots

Informants, like spies, work for two masters. Unlike spies of the James Bond variety, they are unsavory characters, no matter how physically attractive. Because of their lack of trustworthiness, their handlers do well to be suspicious. The problem is that the smallest set may not capture all the liars, so maybe the handlers need to be more suspicious. Whom would you put to the test?

Part VIII

Games

33

The Coach Who Would Do Anything

The two judges happened to be among the world's best women table tennis players. I Fei and Tiffany had each won a recent world championship, but their manner showed only concern for their sport. It seems that Strategy was trying to subvert Talent. I Fei began:

"Two international teams are to meet in either a mini- or maxi-competition at a table tennis tournament. Each team consists of 25 players, but for the minicompetition, only the nine best players will appear. Team A is the better team on average, but team B has some excellent players as well.

"Since you are a mathematician, Dr. Ecco, we will state the problem in a way that you will find clear. Suppose we number the players from 0, the best, to 24, the worst, in each team, so A[0] is the best player on the A team and B[0] is the best player on the B team. Then player $A[i]$ will beat player $B[i]$ by two points for every i between 0 and 24 inclusive; however $B[i]$ can beat $A[i+1]$ for every i between 0 and 23 by two points. Moreover, a strict transitivity holds. So, for example, $A[i]$ can beat $A[i+1]$ by four points and $B[i+1]$ by six points. Similarly, $B[i]$ beats $A[i+1]$ by two points, $B[i+1]$ by four points, $A[i+2]$ by six points, and so on. Do you get the picture?"

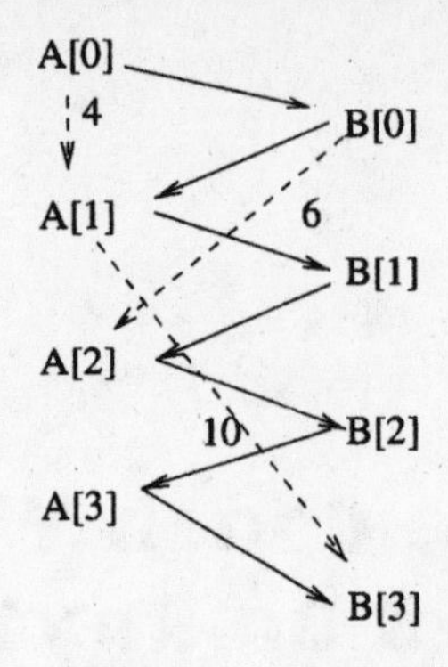

Figure 1. Liane's sketch. Each solid arrow means a win by 2 points. Some of the dashed arrows represent wins by other point spreads. A[0] would beat A[1] by 4 points; B[0] would beat A[2] by 6 points; and A[1] would beat B[3] by 10 points.

Dr. Ecco nodded. Liane started to sketch out the relationships (see Figure 1).

Tiffany continued: "The B team coach, whom we'll call Foxy, knows all of this. The A coach, whom we'll call Trusting Bull, does not. Foxy knows that his team will lose every game by two points if he plays his best player B[0] against Trusting Bull's best player A[0], his second best player B[1] against Trusting Bull's second best A[1], and so on. Foxy therefore wants to figure out a way to organize the competition in such a way that the B team wins more games than the A team, but without arousing the slumbering suspicions of Trusting Bull. Trusting Bull might become suspicious if the average point spread in games when A players win is too high or if the maximum point spread in those games is too high.

"So, Foxy has set himself the problem of pairing up his players against the A team players in such a way that no B player loses by more than 10 and that the average point spread for games when A players win is well under 7. Even Trusting Bull would notice bigger point spreads than this. We are not of course asking for you to help Foxy, but rather whether he can achieve his sneaky goals and, if so, how.

"Here is an example just to show you how this would go: if there were just five players on each side, then Foxy might pair them up as follows:

TEAM A	TEAM B
0	4
1	0
2	1
3	2
4	3

This pairing would enable B players to win by two points in every game except that against A[0]. However, B[4] would then lose by 18 points. Foxy wouldn't choose this strategy, because he would arouse the suspicions of Trusting Bull. Besides, he needs only a majority of games, not four out of five."

1 and 2. Cybernovice: First try to figure out a winning strategy for Foxy in the case of the minicompetition (nine players on each team) and then for the case of the maxicompetition (25 players on each team). In neither case should a B player lose by more than 10 points, and the average B loss should be as near to 6 points as possible. One last wrinkle not mentioned by the champions: In the case of 25 players, at least five of the B wins should be by two points.

Solutions

The naive way to solve this problem is to try all possible orderings of the B players, evaluate each one, and then determine the best ordering for Foxy. This is quite feasible for nine players where the number of orderings is 362,880, but less so (even with quite speedy computers) when the number of orderings is about 15 trillion trillion. So, we observe the pattern of nine and try to extend it.

1. For nine players, the following pairing yields five wins for the B team (against A players 2, 4, 5, 7, and 8 as indicated by the asterisks).

TEAM A	TEAM B
0	0
1	2
*2	1
3	5
*4	3
*5	4
6	8
*7	6
*8	7

The maximum loss for a B player is 10 points (e.g., A[6] vs. B[8]) and the average point spread when a B player loses is 7.

2. For 25 players, the following pairing yields 13 victories for team B (just enough). The maximum loss for a B player is 10 points and the average point spread when a B player loses is $6\frac{1}{3}$. The B team wins at 0, 15, 16, 23, and 24 are all by two points.

TEAM A	TEAM B
0	0
1	2
*2	1
3	5
*4	3
*5	4
6	8
*7	6
*8	7
9	11
*10	9
*11	10
12	14
*13	12
*14	13
15	15

TEAM A	TEAM B
16	16
17	19
*18	17
*19	18
20	22
*21	20
*22	21
23	23
24	24

Inspiration and Offshoots

As journalists have shown in the past, the desire to win can easily corrupt players (who might bet that their team will lose) and coaches (who might arrange to spy on other teams) alike. The remarkable aspect of this particular dishonesty is that it requires neither spying nor bribery nor trick equipment—just organization. Is there a mathematical way to detect such a cheat? I really don't know.

34

The Nimmerics of Washington

I was just figuring out how I could force a draw in our chess game when the doorbell rang. It rang again almost immediately. Then again. "Must be the Feds," Ecco said with a sigh. "No patience at all." Ecco's 10-year-old niece, Liane, who had been watching our game with an ironic smirk, said "I'll get it," and she ran toward the door.

Federal Agent Thomas flashed his ID badge to all three of us, took a seat without being invited, opened his briefcase, pulled out a folder, and began to speak. "We were keeping Wundermann under observation. Our agents saw him yesterday morning. By the afternoon, he was gone. All he did was leave this note. The trouble is that we don't know what to make of it."

Wim Wundermann was a professional colleague of Ecco's. He had dropped out of sight a few months earlier, saying that he had discovered patterns he called Nimmerics. Ecco told him that he should steer clear of applications to encryption.

"Well, what does the note say?" Liane asked.

Agent Thomas showed us the paper. We all noticed that the writing was strange, especially at the beginning. " 'This is useful information for any Nim-Wise player. You know how to Play this Easy game of Nim Though it is not as easy as it seems.' "

"Oh, yes, I remember," Liane interrupted. "You start with a bunch of stones and you can take one, two, or three of them in your turn. You win if you remove the last stone."

"Right," said Ecco, "and do you remember the winning strategy?"

"Let's see," Liane said. "You try to end each turn leaving 0, 4, 8, 12, 16, ... [any multiple of 4] stones to the other player. Then you force the number of stones to exactly four less at the end of your next turn."

"What would happen, Liane," Ecco asked, "if each player could remove one more stone than the last player? So, if the first player could remove up to three in the first move, then the second could up to four in the second move, and the first could remove up to five in the third move, then the second could remove up to six in the fourth move, and so on?"

"Hmmm," Liane paused for a moment. "The second player would still win if there were four stones at the beginning, but would even win if there were five, because he'd be able to take up to four in the second move. I'll have to think about this some more."

"May I continue?" Thomas demanded. "We are talking about a disappearance here of a major national asset and ... "

"Yes, I'm so sorry," Ecco apologized (though I thought I detected a trace of mockery in his voice). "Please go on."

Thomas read on, but in a louder, slower voice. " 'I will give you two puzzles. Suppose I present a bowl containing one $1 bill, two $2 bills, three $5 bills, and four $10 bills [Imagine a bowl having: 1, 2, 2, 5, 5, 5, 10, 10, 10, 10]. Play consists of moving a single bill at a time from the bowl to the table, which has no bills initially, until the amount on the table reaches or exceeds a given target number. If a player's move makes the amount reach that target, then that player wins. If a player's move makes the amount exceed that target, however, then that player loses. For example, if the target number is 3, then the second player wins, because the first player will either exceed $3 on his first move or will place a $1 or $2 bill on the table in which case the second player will make the amount equal $3 on her move.' "

"Yes, but the first player usually has an advantage," Liane observed. Thomas glared at her, but Ecco looked at her curiously.

"Oh, it just feels that way," she said with a shrug. "It's true for Nim."

Ecco nodded, but then turned his eyes to Thomas. Thomas read with even more deliberation. " 'Let *x* be the first target number 29 or greater in which the second player can force a win. Let *y* be the first target number 41 or greater in which the first player can force a win. Find me at that location.' "

1. *Cybernovice: Can you find Wundermann's address at x and y?*
2. *Advanced cybernovice: Does either x or y change if there are twice as many 1s, 2s, 5s, and 10s? If so, to what?*

Ecco nodded thoughtfully, then said, "OK, so let's play Wundermann's game, Miss Feels-That-Way."

While Thomas paced the room, Ecco worked out the problem by playing the game with Liane. After he gained enough intuition, he gave Thomas the answer. "These two numbers are the *x* and *y* you are to look for, perhaps they are street and avenue addresses. Liane has a theory about the letters."

"NWPENT," she said. "Look at the beginning of the note. Those are the weird uppercase letters. The others are there by grammar."

Federal agent Thomas stared at her as if puzzled for a moment, then his face cleared. He collected his papers, stood up, and left without a word.

A few days later Thomas was back. "We found the place. A penthouse apartment in the Northwest district near the street corner you identified," Thomas said, nodding to Liane and Ecco. "Now, Wundermann has left us with a new puzzle. The guy's insane.

" 'Consider a variation of Nim which I'll call Expanding Nim. In that variation . . .' "

That new game turned out to be exactly the one Ecco had asked Liane about earlier. The possibilities expand at each move, making this a difficult game to play.

"'Call a number N a WinTwo number if the second player can force a win when there are N stones to begin with. Find the three smallest Win2 numbers 59 or greater.'"

"Well, Liane, can you do it?"

"Sure," she responded cheerfully. "Here they are. . . ."

3. Cyberexpert: Can you find Liane's solution?

Solutions

1. $x = 30$, $y = 43$. A way to solve this is to construct a game tree in which each node represents the state of the game (a set of bills) and each child node corresponds to the removal of a bill. The tree can become large, but a standard technique called alpha-beta pruning can shrink the size of the tree. This saves time.

2. If there are twice as many bills in the container, then $x = 29$ and $y = 41$.

3. For Expanding Nim, notice that 4 and 5 are both Win2 numbers since the second player can remove the last stone no matter how many the first player removes in his turn. For example, if there are five and the first player removes one stone, then the second can remove four stones. Similarly, 10, 11, and 12 are all Win2 numbers, since the second player can force a situation in which the first player is left with either six or seven stones at his second move. For example, if there are 10 stones to begin with and the first player removes 3, then the second player removes 1 and the first player is faced with 6. Since the first player at this point can take only five, the second player will win.

So, the following are all Win2 numbers:

$\{4, 5\}$ (i.e., 4 or 5);

$\{4, 5\} + \{6, 7\}$ (i.e., 10, 11, or 12; which are the sum of $4 + 6$, $5 + 6$ [or $4 + 7$], and $5 + 7$ respectively);

$\{4, 5\} + \{6, 7\} + \{8, 9\}$ (i.e., 18 through 21);

$\{4, 5\} + \{6, 7\} + \{8, 9\} + \{10, 11\}$ (i.e., 28 through 32);

$\{4, 5\} + \{6, 7\} + \{8, 9\} + \{10, 11\} + \{12, 13\}$ (i.e., 40 through 45);

$\{4, 5\} + \{6, 7\} + \{8, 9\} + \{10, 11\} + \{12, 13\} + \{14, 15\}$ (i.e., 54 through 60);

$\{4, 5\} + \{6, 7\} + \{8, 9\} + \{10, 11\} + \{12, 13\} + \{14, 15\} + \{16, 17\}$ (i.e., 70 through 77).

So, Liane's answer is 59, 60, and 70.

The winning strategy for the second player goes like this for, say, 71. In the first two moves, force the number down to

$$\{6, 7\} + \{8, 9\} + \{10, 11\} + \{12, 13\} + \{14, 15\} + \{16, 17\},$$

that is, 66 (taking the sum of the minimum of each pair) through 72 (taking the sum of the maximum of each pair). In the second two moves, force the number down to

$$\{8, 9\} + \{10, 11\} + \{12, 13\} + \{14, 15\} + \{16, 17\},$$

that is, 60 through 65. Then 52 through 56, 42 through 45, 30 through 32, 16 or 17, then 0.

Inspiration and Offshoots

The reason Nim, or even the substantially more complicated games of the puzzle, are feasible to solve is that each player has complete information and the winning strategy can be computed fast (chess, by contrast, enjoys complete information but requires exponential search). An interesting variation would be a kind of probabilistic Nim in which the first moves are based on chance and the later moves are not. Here is a first example: Suppose you start with 30 objects always. The first two turns are based on the throw of a single die. The player who moves can choose any number of objects from 1 to the number on the die (so if the die has the value 4, then the player can remove one, two, three, or four objects). After the players both complete their first move, the first player retains the choice presented by his roll and similarly for the second player. The second player should have the better odds, I think. What do you think?

35

Baskerhound's Simplest Game

Benjamin Baskerhound himself was at our door. Kidnapper of Ecco, stealer of submarines, would-be founder of a colony in Antarctica, Baskerhound had promised the authorities that he would avoid mischief ever since his radical brain surgery and subsequent (in my opinion, ill-advised) presidential pardon. "I now invent fictional strategy games," he said, paused, his eyes twinkling, "without ulterior motive." He had lost weight since the last time I saw him. He now would have blended into the professorial class of any college campus: sports jacket, unruly hair, crooked glasses, and intelligent, bemused eyes. I still saw wickedness behind those eyes, I must confess.

"I have invented a game now that I can't solve. It's so easy to explain, I call it 'Simple.' "

I saw Liane rub her hands in delight. Ecco too leaned forward in his chair. In spite of (or because of?) their history together, Ecco seemed comfortable with Baskerhound.

"The game is played on a grid without boundaries," Baskerhound began. "Players—we'll call them player x and player o—alternate moves as in tic-tac-toe. Player x goes first. The object is to get four-in-a-row. But the four must be connected either vertically or horizontally. Diagonal foursomes don't count."

"Still, as in tic-tac-toe, the first player has quite an advantage," Liane observed. "Is there any compensating advantage for the second player?"

"Well, yes," said Baskerhound, "but that is where I need your help. To win, player x (the first player) must get four-in-a-row before player o does, and player x must win within his first 10 moves. If after 10 moves of player x, that player has no vertical or horizontal four-in-a-row, then player o wins."

"So, no game can have more than 19 x's and o's on the page," said Liane, showing off a little.

"You do credit to your Uncle Ecco here, young lady," Baskerhound said to Liane. "But the real question is: does either side have a winning strategy? If so, is 10 too low a number or doesn't it matter?"

After a few seconds of silence, Liane spoke up. "If only three in a row were necessary, then player x could win in exactly three moves. Here is how:

"Player x moves and o moves next to x

x o

"Now, player x moves orthogonally to where o moved:

x o
x

"Since x needs only to get three in a row, there is no way for o to stop x."

"Right," said Ecco. "The question of four should bend to the same reasoning. After all, any three in a row with open ends on both sides should force a win. Still, the problem is not trivial at all. We need to think."

Ecco invited Baskerhound and Liane to a table. After an hour of playing on paper and chessboard, with much chuckling and exclamations of surprise, Ecco, Baskerhound, and Liane turned to me with a look of triumph.

"We think we have it," said Liane.

1. Cybernovice: Does either player have a winning strategy and, if so, what is the strategy? If player x can force a win, then how many moves are needed? If player o can force a win (by preventing player x from winning in 10 moves), then does x ever have a winning strategy?

"But there are some open questions," Ecco added. "We also have some variants of the game. Can either player be sure to get four-in-a-row more than once? What is the situation if winning requires five-in-a-row?"

2. Cybernovice: What do you think?

"In the four-in-a-row game, how soon or ever can the first player force a win, if that player must always (after player x's first move) place a mark that is one square away (either horizontally or vertically) from some other mark? That is, after player x's first move, every new x must be next to some o or x."

I don't know the answers to the question, nor do I know whether my friends—or even Baskerhound—ever figured it out.

3. Cybernovice: Can you answer Dr. Ecco's questions?

Solutions

1. Player x can force a win in the original game. Here is how. Without loss of generality, x can force the following situation (or something symmetric):

```
x o
x
```

Now, o has two choices (because any other would lead to three x's in a line and make it impossible for o to stop the result):

```
o            x o
x o    or    x
x            o
(a)          (b)
```

Let's follow (a) first. Player x forces player o by threatening a win:

```
o
x o
x
x
```

Player o stops the certain win:

```
o
x o
x
x
o
```

Now player x plays under the first o:

```
o
x o
x x
x
o
```

Now, o must stop the two x's and has two choices:

```
o             o
x o           x o
x x o   or  o x x
x             x
o             o

(a′)         (a″)
```

From (a′), we get to

```
  o
  x o
x x x o
  x
  o
```

followed (forced by):

```
    o
    x o
o x x x o
    x
    o
```

and this is followed by

```
    o
    x o
o x x x o
  x x
    o
```

which leads to two forced wins (a forced win is three x's in a row without o's). So, this finishes case (a′).

From (a″), player x can play a forcing move:

```
  o
  x o
o x x x
  x
  o
```

So, player o responds with:

```
  o
  x o
o x x x o
  x
```

Now, player x does:

```
  o
  x o
o x x x o
  x   x
  o
```

and can force a win.

Now let's look at (b). Player x plays a forcing move

```
x
x o
x
o
```

So, player o must respond:

```
o
x
x o
x
o
```

Now, player x does:

```
o
x
x o
x x
o
```

So, player o must respond with either:

```
o             o
x             x
x o    or     x o
x x o       o x x
o             o

 (b′)        (b″)
```

In case (b′), player x plays a forcing move and player o responds:

```
  o
  x
  x o
o x x x o
  o
```

Then player x plays:

```
  o
x x
  x o
o x x x o
  o
```

and guarantees a win. This takes care of case (b′).

In case (b″), player x plays a forcing move and player o responds:

```
  o
  x
  x o
o x x x o
  o
```

Now, player x plays the following clever move (unconnected to any other node):

```
  o
  x   x
  x o
o x x x o
  o
```

This ensures a win by x and takes care of case (b″).

2. Ted Alper showed by use of a clever tessellation that there is no winning strategy for five-in-a-row. Here is the tessellation.

```
A B C C
A B D D
E E G H
F F G H
```

Alper puts in this way: "Two adjacent squares are adjoined to form a "domino"... and two adjacent dominos share the same orientation (whether horizontal or vertical), but the next two dominos in any direction from them will have the *other* orientation. Any winning position of length 5 must have 2 squares from the same domino. When the first player takes any square, the O player takes the remaining square," thus ensuring eternal frustration to the X player. Victor Allis and his colleagues showed in the computer conference AAAI 1993 that the first player has a forced win if diagonals are allowed.

Inspiration and Offshoots

Have you ever been in a restaurant where you find yourself waiting too long for the food? The people you are with are smart but quiet. You can play checkers with the sugar bags, but you want a more challenging game played with relatively few bags. Simple would work. You need only 16 bags. If your friends discover the forced win, ask them to work on the unsolved touching variant.

36

Wordsnakes

"The graphic artists say they are ready," our visitor said with delight. Mike Johnson was the games editor of a big city newspaper that many coffee and latte shop denizens have been seen to read in cities and towns everywhere. "They have a design of a long sequence of letters that twists along a page and even bridges over itself. You see our solutions will look like that and it is important that they, well, fit.

"Let me explain the game. Given a collection of words, a word-snake is a list of those words without repeats such that some suffix of each word is a nonempty prefix of the next word in the list. [Suffixes of the last word are unconstrained.] The score of such a consecutive match is the square of the number of letters in the overlap.

"For example, the words *house* and *sea* have an overlap of two letters, hence a score of two squared or 4, in the given order because the suffix *se* of *house* is the prefix of *sea*. On the other hand *beret* and *timber* have an overlap of one, and score of 1, in the given order because of the letter *t* but have an overlap of three, and score of 9, in the reverse order because of the letters *ber*. So, some wordsnakes have higher scores than others.

"I spoke about the graphic artists because we will present the wordsnake as a long word having a nonredundant representation of the overlap. We call this 'the wordsnake in long form.' This will give us *housea* for *house* and *sea* and will give us *beretimber* for *beret* and *timber* in that order but *timberet* in the opposite order. If the initial collection is long, then the wordsnake in long form can be quite lengthy too."

"Cool game!" 12-year-old Liane said. "Give us a try."

"I had hoped you would have that reaction," Johnson said. "Here is a list of words in alphabetical order. Your job is to rearrange the words into a wordsnake that gives the highest score possible. No need for you to lay it out in long form. Our artists will do that."

alas	eternal	merger	terrestrial
ally	geriatric	pediatric	terrible
blemish	gerund	penultimate	tessellate
blend	icer	rates	trials
certain	incredible	sea	trice
dense	ingrate	seem	tricky
denude	invent	shape	underdevelop
eloped	lasting	stinger	yes
emerge	later	sudden	yet
essential	lending	tense	

1. Cybernovice: Liane found a list with a score of 340 and a length in long form of 145. Can you do at least as well scorewise?

"I knew my niece would find a solution, Mr. Johnson," Ecco said after Liane had presented her solution, including a design for the wordsnake in long form (she couldn't resist). "We have been working on Hamiltonian paths together lately. In the meantime, I have some variants to ask you about: One can imagine trying to find the longest possible wordsnake in long form, in which case *beretimber* would beat *timberet*, or the shortest possible wordsnake in long form, this need not always have the highest score, though often it will, for this list. Finally, I think there could be a fascinating competition in which one tries to design a wordsnake whose representation in long form would be 50 letters or fewer and whose score would be as high as possible. Such a competition could be conducted in any one of several languages."

Johnson promised to think about these challenges but said he wanted to present the game as it was to his editor-in-chief. Ecco and Liane worked on the 50-letter variant for a while, but remained unsatisfied with their solutions. They did, however, find a list whose 18-letter long form yielded a score of 338.

2. Cyberexpert: Can you do at least as well as Dr. Ecco and Liane for 18 letters? The best solution I know of in English so far has a value of 663, though there is also a solution with a value of 843 using only 15 letters. There is a German solution having 18 letters with a remarkable score of 993.

Solutions

1. Here is Ecco's ordered list. This wordsnake has a score of 340.

invent
tense
seem
emerge
merger
geriatric
tricky
yes
essential
ally
yet
eternal
alas
lasting
stinger
gerund
underdevelop

eloped
pediatric
trice
icer
certain
incredible
blemish
shape
penultimate
terrible
blend
lending
ingrate
rates
tessellate
later
terrestrial
trials
sudden
denude
dense
sea

Patrick Schonfeld gave a succinct explanation of the method: (i) Each word in the given 39-word list has at least one successor; (ii) for each word in the list, determine which successors have the maximum overlap with the word; and (iii) try to use those with maximum overlap.

Here is the wordsnake in long form.

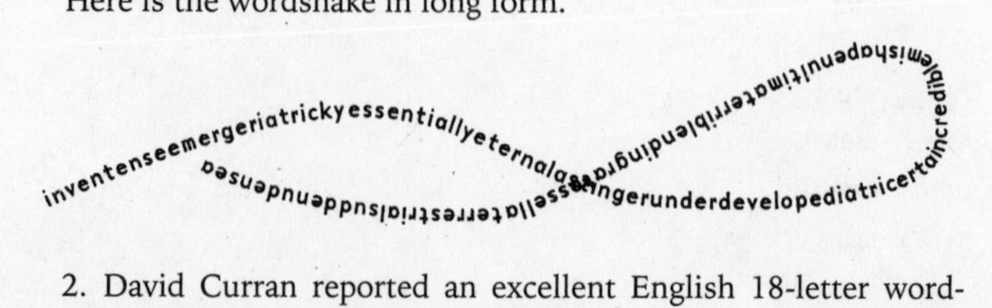

2. David Curran reported an excellent English 18-letter word-snake: "conversationalists." This simple word yields a score of 663

having the breakdown: con, conversation, conversational, conversationalist, conversationalists, and lists.

Genny Engel managed to beat this with the 15-letter wordsnake "misrepresentions": misrepresent, misrepresentation, misrepresentations, representation, representations, presentations, and ions. This gave her a score of 843.

Martin Laeuter of the mathematics department at Leipzig suggests the following 18-letter German wordsnake: verwirtschaftetest. From this, the following words are present, giving a total of 993 points, an impressive achievement: verwirtschaftete, verwirtschaftetes, erwirtschaftetes, erwirtschaftetest, and wirtschaftetest.

For those who don't know German, here is an explanation: "OK, now to the meanings. The used verbs are verwirtschaften, erwirtschaften, and wirtschaften. To begin from the last, wirtschaften is the action of organising things economically, e.g., making ends meet in the household or the tasks of a manager (to manage affairs). erwirtschaften is working so that money or other means are available to pay expenses. verwirtschaften is simply bad wirtschaften, bad management."

Inspiration and Offshoots

If you are good at crosswords, you will recognize the virtuosity of starting a puzzle at some arbitrary place and then filling in words in such a way that the next word always includes a letter from a previously entered word. Trying to achieve a large overlap among words in a list shares a similar motivation. The two can be put together: Find a two-dimensional structure of the words (so words can be read from left to right or top to bottom) such that words can overlap as in the puzzle or can use single letters in other words by crossing those other words. Scoring is the same as in the puzzle. It is an open problem to design the best crossword by that criterion for the words in this list.

Ciphertexts for the Time: A Bonus Puzzle

You may have noticed that the puzzles are not in chronological order. This puzzle mixes decryption with reordering. Associated with each chapter number is a single line of ciphertext as you can see. First, reorder the chapters into approximate chronological order, then decode the manifesto . . . if you can. You'll find that the decryption effort informs the ordering effort and vice versa in a symbiotic way. I'm not presenting the solution for this one. You'll know when you get it.

1: z achmfsq xhpv ml zuvh xlw
2: igowfdqx ct c wxgtdsqdui pqkuiq,
3: f wxrzqt maigt yjalt gtlvkxbit.
4: xcdsqx dsch ct c tfjtdudfdugh vgx
5: xii yldncptalo xfxpixeil.
6: szfv shkexzme lc ixv hozsziq il ybvhf.
7: bqosv ozuovsofgo mdrb ho vsgq ofedxq be nvap duef
8: nbcrpnbcths bgo hfnvecbctfg.
9: ablmztzg dm dc, d'x
10: bykfqa zcla wywoft?

11: nahek whh uobjn ap qradhkin, znobt
12: uewcrpw, fngtrpewtscs srfemo jtpq crp
13: yzxkzxe ajs j qbew ba anurbog j dmllux
14: zi zy h kmzdkvsq xkthm hozsziq, fzmu la
15: yzxo bk qmbuwa j oxak nmk nc mocjvbubje
16: ebtqaovohcblb aksy fklb fnp wkmyb, rnl
17: scefardezkrai sudl dkm dgjnfaq.
18: jadn ipvl x onprmsldv bc xc
19: xgfforu vi etvu khhb vilholr urdyjtvia
20: etvibvia. stryrdu pdim hw etr
21: petcbdvm ydl qdgmdz dmn
22: fuzihbu e pezbs vyeks, qivf zsahwzs
23: apza dlerpncurqaq lcqa gn wnenuzkrqaq.
24: cb lakl enmmvy-bovscph of bcwevt
25: ra rq mu. nffd'q gnkrns (zem r zwunn orap prl)
26: iwly d'x qpm l glm.
27: afalexj afn gemmef rofro ar poll ar
28: oji riwg iwt xrfcxkfgxjk fks latfgxnxgo
29: bdwdcnd qpjydl, yd ckdqombtqpz kbpz
30: zybxh glt hwul gxe hwu mksajhub hk tk hwu dbjlh vkbe.
31: jm vwzafn in whcwdauwnj jm vwjwduanw
32: mk ar nmk on w qkrnab lma cwb
33: vjkxebjua ne wnbog anvxkzbog xuax.
34: lmwx on wb aibomkzronx?
35: xez fii qhku kiqyhfefwij fwqs xiz
36: g tuua cjuxhykl gnkjh ajiirux yx vwuhwub

About the Author

Dennis E. Shasha is a professor at New York University's Courant Institute where he does research on biological pattern discovery, combinatorial pattern matching on trees and graphs, database tuning, and the fast analysis of time series. After graduating from Yale in 1977, he worked for IBM designing circuits and microcode for the 3090 and completed his Ph.D. at Harvard in 1984. He is the author of five books, including two books about the mathematical detective Dr. Ecco entitled *The Puzzling Adventures of Dr. Ecco* (1988) and *Codes, Puzzles, and Conspiracy* (1992) and the coauthor of a book of biographies about great computer scientists called *Out of Their Minds: The Lives and Discoveries of 15 Great Computer Scientists* (1995). He currently writes a monthly puzzle column for *Scientific American.*